BACTERIOLOGY METHODS FOR THE STUDY OF INFECTIOUS DISEASES

BACTERIOLOGY METHODS FOR THE STUDY OF INFECTIOUS DISEASES

ROWENA JENKINS
Department of Microbiology and Infectious Disease
Swansea University Medical School
Swansea University, UK

SARAH MADDOCKS
Department of Biomedical Sciences
Cardiff School of Sport and Health Sciences
Cardiff Metropolitan University, UK

ELSEVIER

ACADEMIC PRESS
An imprint of Elsevier

Academic Press is an imprint of Elsevier
125 London Wall, London EC2Y 5AS, United Kingdom
525 B Street, Suite 1650, San Diego, CA 92101, United States
50 Hampshire Street, 5th Floor, Cambridge, MA 02139, United States
The Boulevard, Langford Lane, Kidlington, Oxford OX5 1GB, United Kingdom

Notices
Knowledge and best practice in this field are constantly changing. As new research and experience broaden our understanding, changes in research methods, professional practices, or medical treatment may become necessary.

Practitioners and researchers must always rely on their own experience and knowledge in evaluating and using any information, methods, compounds, or experiments described herein. In using such information or methods they should be mindful of their own safety and the safety of others, including parties for whom they have a professional responsibility.

To the fullest extent of the law, neither the Publisher nor the authors, contributors, or editors, assume any liability for any injury and/or damage to persons or property as a matter of products liability, negligence or otherwise, or from any use or operation of any methods, products, instructions, or ideas contained in the material herein.

Library of Congress Cataloging-in-Publication Data
A catalog record for this book is available from the Library of Congress

British Library Cataloguing-in-Publication Data
A catalogue record for this book is available from the British Library

ISBN: 978-0-12-815222-5

For information on all Academic Press publications visit our website at https://www.elsevier.com/books-and-journals

Publisher: Andre Wolff
Acquisition Editor: Linda Versteeg-buschman
Editorial Project Manager: Carlos Rodriguez
Production Project Manager: Poulouse Joseph
Cover Designer: Greg Harris

Typeset by TNQ Technologies

Contents

Acknowledgement

The authors would like to thank Dr Aled Roberts, Leighton Jenkins and Phillip Butterick for their support during the preparation of this book, in particular for their help in preparing materials for imaging.

CHAPTER ONE

Fundamental skills for infectious disease research

Contents

1.1 Introduction

When planning any experiment, it is important to make sure that all of the equipment and reagents are pre-prepared and arranged within a sensible working area so that disruption of your work is minimal. This includes considering whether ice is available for reagents that need to be cold and somewhere to dispose of tips, glassware, and waste solutions, for example. This chapter will help you to consider the risks that need to be assessed before you start and what you need to set up to carry out experiments you are planning. This information will be applicable irrespective of the type of experiment you do.

1.1.1 Health and safety

Before you can start in the laboratory, it is essential that you take some time to consider procedures that might need to be in place to allow you to work safely.

When working within the microbiology laboratory, you are going to be handling microorganisms. It is important for you to know what kind of risk they might pose to you and anyone else working within the laboratory. Because the microbial cultures you are going to be handling can often contain many millions of potentially infectious cells, guidelines are in place to help you handle them safely. The Advisory Committee on Dangerous

Bacteriology Methods for the Study of Infectious Diseases
ISBN 978-0-12-815222-5
https://doi.org/10.1016/B978-0-12-815222-5.00001-8

Table 1.1 Hazard group definitions according to the Advisory Committee on Dangerous Pathogens and examples of microorganisms that come under those categories.

Hazard group	1	2	3	4
Definition	Unlikely to cause human disease	Can cause human disease and may be a hazard to employees; unlikely to spread to community; effective prophylaxis or treatment is usually available	Can cause severe human disease and may be serious hazard to employees; may spread to community, but effective prophylaxis or treatment is usually available	Causes severe human disease and is serious hazard to employees; likely to spread to community and no effective prophylaxis or treatment is usually available
Example microorganisms	*Bacillus subtilis*, *Saccharomyces cerevisiae*	*Bacillus cereus*, *Campylobacter jejuni*, *Pseudomonas aeruginosa*, *Staphylococcus aureus*, Epstein–Barr virus	*Mycobacterium bovis*, *Coccidioides posadasii*, Borna disease virus, West Nile fever virus	Marburg virus, Sudan ebolavirus

Pathogens provides an approved list of biological agents (microorganisms) as referred to in Control of Substances Hazardous to Health (COSHH), which includes bacteria, fungi and viruses that could pose a risk to human health. You should refer to this if you are uncertain about the hazard rating of the organism with which you are working. Broadly, the biological agents are split into four categories (Table 1.1), in which Category 1 agents present no risk to human health, and Category 4 the highest risk to human health. The categories are sorted into these groups based on their likelihood of causing disease, ability to spread to the community, and availability of treatment or prophylaxis. You will need to know what category your organism falls into before you start work, because it is necessary to fill out a COSHH form to ensure you have control measures in place that are appropriate to the level of risk to which you going to be exposed.

Organisms from Categories 1 and 2 are most commonly employed in laboratories because they can be used either on a bench top or within a laminar flow hood. Each laboratory needs to have the appropriate containment level rating to handle the corresponding organisms. For example, a Hazard 2 agent can be handled only in a lab that is set up to at least Containment Level 2 standards. The containment level is normally concerned with the structure and layout of the laboratories, the type of equipment required (such as autoclaves) and the internal procedure in place to manage facilities and processes such as waste disposal. It is likely that all of this will already be well-established in most laboratories. Organisms from Categories 3 and 4 require more specialist handling and laboratory setup; fewer laboratories are set up to handle organisms from those categories. For more information on the classification system, you can go to the Health and Safety Executive (HSE) biosafety webpages and consult the Advisory Committee on Dangerous Pathogens documents. If your susceptibility to infection is likely to be altered owing to a pre-existing health condition, medication, reduced immunity or pregnancy/breastfeeding, you should reassess the risks involved because the hazard ratings are assigned without taking those factors into consideration.

There are further considerations if you are planning to work with genetically altered microorganisms, because this is also not accounted for in the hazard rating and could alter the risk associated with handling them. Regulations exist governing the use of genetically modified organisms (GMOs); information on these regulations may be found on the HSE webpages. If you are starting out in microbiology, it is unlikely that you would be responsible for applying for a licence to work with these organisms. The HSE would need to be informed the first time a laboratory intended to work with a GMO, so it is worth checking before you start whether the laboratory in which you will be based is already working with GMOs and has notified the HSE of that work.

When conducting an experiment within a microbiology laboratory, it is likely that in addition to handling microorganisms, you will need to use a range of chemicals and reagents to achieve your desired outcome, and that you will not be immediately familiar with many of these chemicals. To prepare them safely, it is important that you check the safety data sheet (SDS) for each chemical. The SDS contains the information you need to work safely with any particular substance. It lists a variety of information such as hazard identification, toxicity, reactivity and stability of substances, as well as how to store, dispose, transport and deal with spills of that substance. The SDS information will also cover what first aid measures need to be taken when handling the substance. For more information on SDS, you can look on the HSE website (http://www.hse.gov.uk/index.htm); for sample SDS sheets, go to the manufacturer's website (for example, Sigma–Aldrich or Thermo Scientific) for the chemical you are purchasing. Most will have a link to the SDS sheet for that chemical.

When you are purchasing chemicals, SDS information is often available online on the manufacturer's website. It is important that you read it thoroughly for each reagent or chemical that you work with and that you use that information to fill in your risk assessment for that experiment. You may find that your laboratory already has an inventory of SDS sheets for certain chemicals to which you can refer in the first instance. However, it is important to check that none of the information has changed, because risk and handling recommendations do so from time to time, and the risk assessment may need to be updated accordingly. It is useful to include the most important points about substances you are handling on your risk assessment and then keep the risk assessment on your bench where other others who use the laboratory can find it. This means that if there is an accident or spill while you are not there, information about how to deal with it will be readily available to others.

There are a number of different hazards to which you might be exposed when handling chemicals and reagents. The hazard or hazards associated with each chemical will often be displayed as a pictogram on the container of the chemicals you purchase. These pictograms were developed to provide a Globally Harmonised System which standardised the labelling of hazards associated with chemicals across large parts of the globe. It is imperative that you be able to identify the symbols and the associated risk correctly so that you can handle the chemicals in a safe manner. See the HSE website for pictogram information (http://www.hse.gov.uk/chemical-classification/labelling-packaging/hazard-symbols-hazard-pictograms.htm).

The chemicals that you purchase and work with may have none, one, or more of these of these pictograms (Fig. 1.1). Care should be taken to check for this information and to handle the chemical or reagent appropriately.

Information provided on the SDS will help you and your colleagues to remain safe in the laboratory. A good way to record the information and risk is to complete a COSHH or risk assessment form. A risk assessment of all of the procedures you plan to carry out will help you to consider any risks you may encounter when working and will reduce the likelihood of accidents occurring by implementing suitable control measures. There are many different risk assessment forms. You will probably find your place of work already has a template in place that will help you to identify any significant risks within the processes you are going to carry out, who might be

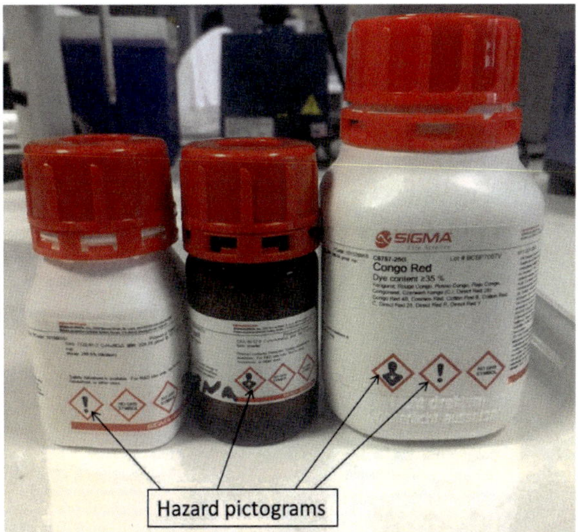

Figure 1.1 Example of labelling you might see when handling chemicals in the laboratory.

affected by those risks (just you or others within the laboratory) and which control measures are already in place (many laboratories have training and standard operating procedures in place, as well as suitable work spaces and protective equipment). It will also allow you to consider how to minimise risk by implementing further control measures if needed.

Within the risk assessment form, you need to consider whether you are working with anything hazardous (you may not be). Such hazards could be chemical or physical things including hot media coming out of the autoclave or the use of a −80°C freezer. You need to think about how processes involved in your work might harm you or others and how you can mitigate the likelihood of them causing harm. For example, if you were planning to use a fine powder that could cause dust inhalation, could it be swapped for a pelleted form? Generally, if it is a hazardous substance and there is no way to replace it, you should consider how to minimise its use. Check the route of exposure that will cause harm: for instance, is it likely to be inhaled and irritate your respiratory tract or does skin contact cause the problem? Then, you can decide the best way to reduce exposure, such as by using a fume hood. Once you have considered all of the risks and put your mitigation measures in place, you will probably need to share your risk assessment form with either your supervisor or your laboratory manager (see an example risk assessment in Fig. 1.2). Once your assessment has been signed off, you can move on with your experiments with confidence, knowing that you

RISK ASSESSMENT FORM

Name:	Dr Rachel Jones
Work to be undertaken:	Antibiotic sensitivity testing of *staphylococcus aureus*
Laboratory location	Lab 1

This assessment should be filled in using information available from safety data sheets and reviewed at any point when it is thought that the original assessment might no longer effectively cover the work being carried out. This could be due to a change in experimental design, substances in use or changes to the hazard rating/guidance of a substance in use.

When categorising the risk it is important to consider the consequence of exposure to the hazard itself. Could the hazard cause an infection in one individual or could it cause fatalities in a number of people. Then consider the likelihood of that outcome occuring (is it likely or remote that harm would occur with the control measures that are in place). You can then use the table below to approximate the risk of the procedure. If the risk is moderate to high you will need to considre further risk control.

Hazard consequence	Likelihood			
	High	**Medium**	**Low**	**None**
Severe	High	High	Medium	None
Medium	High	Medium	Medium/Low	None
Low	Medium/Low	Low	Low	None
Negligible	None	None	None	None

Risk Level	
Rating	**Action**
High	High level of risk, do not proceed unless control measures Can be implemented to bring risk to an acceptable level.
Medium	Moderate level of risk, risk control measures to be fully Considered when completing risk assessment
Low None	Low level of risk, monitor and review risk on a regular basis No action required

Procedure to be undertaken(quantity/concentration of substances used, how long procedure takes): e.g. Determining the antibiotic sensitivity of *staphylococcus aureus* to tetracycline hydrochloride 512 μg/ml – 0.5 μg/ml made from powdered stock.

Person at risk (e.g staff/student)	Staff
Substance being assessed	Tetracycline

Classification (GSH for chemicals or ACDP for organisms):

Hazard statements: H361 suspected of damaging fertility or the unborn child. H411 Toxic to aquatic life with long lasting effects.

Hazard type:
Dust and liquid

Exposure route:
Ingestion/skin

Personal Protective Equipment required:
Gloves, eyewear, laboratory coat

Control measures already in place (including disposal and spills):
Eye protection, gloves and lab coat to be worn when handling substance. Use in awell-ventilated area. In case of contact with eyes flush well with water, eyewash to be available. If skin contact made, wash with soap and water. If ingested contact a physician.
All materials in contact with biological samples will be autoclaved before disposal. Spills will be cleared up with paper towels before being autoclaved.

Risk rating (after controls are in place):
Low

Any new controls needed to achieve adequate risk control (list below):
Risk assessment to be reviewed on yearly basis or when any new guidance is given on the substance being used No further controls needed at this stage.

Date assessed: .... Title of activity: ....

Assessed by: ________ Checked by: ________

Signature: ________ Signature: ________

E-mail address: ________ E-mail address: ________

When reviewed add the date of the review to the form and note if any control measures were changed. People working with the substance must then sign and date to show that they aware of the changes made.

Figure 1.2 Sample Control of Substances Hazardous to Health form. These vary among workplaces. Although a chemical is used as the example here, you can include microorganisms and physical risks when assessing the risk of the procedure you want to carry out.

can correctly and safely handle all of your substances. In case an accident occurs, most laboratories have an accident and near miss book that can be filled out once the accident has been dealt with. This can inform future risk assessments and improve best practice within the laboratory.

Having carried out your experiments successfully (hopefully), you will need to dispose of the materials with which you have worked. Safe disposal should be included within the risk assessment form. When working with bacteria, this can involve several different routes of disposal, depending on what you have been doing and whether you are using glass, plastic or sharp materials. For this book, we will discuss only requirements in place for experiments involving hazard Category 1 and 2 microorganisms, because laboratories working with hazard Categories 3 and 4 have more complex arrangements in place.

When disposing of agar and broth media, it is usual to place plastic materials (tips/agar plates/disposable inoculating loops, etc.) inside an autoclave bag (some places use double bags to reduce the risk for leaks) held within a bin or bucket that can be autoclaved, whereas non-sharp glassware (conical flasks/universal containers/media bottle) go into a bin or bucket without the autoclave bag. These containers will then be autoclaved at 121°C for 20 min to sterilise the contents. Then the autoclave bag and its contents are disposed of as non-hazardous waste according to local laboratory guidelines. Once the glass waste has been autoclaved, any liquid content can usually be disposed of into the laboratory drainage system, and the glass is washed and placed in a drying cabinet for use the next time. If you have autoclaved agar, it is not a good idea to pour it down the drain because it will solidify. Wait until it has set and dispose of it as a solid waste instead. Some chemicals require special disposal by an external company that will collect the waste. Check with your laboratory manager to see whether this is the case for any chemicals with which you are working. If you are working with glass slides, scalpel blades or any other sharp items, there should be a designated sharps bin for these to be disposed of separately. If you put the glass into normal bins containing autoclave bags, the sharps will pierce the bags and allow the contents to leak out. Always check with your laboratory manager if there is chemical about which you are unsure after you have consulted the SDS.

When you start in a new laboratory it is usual for laboratories to have their own local health and safety guidelines and training, so be prepared to attend a briefing session and read through any relevant documentation you are given. Often you will be required to sign a document that states that you are aware of safety issues and local regulations and will abide by those regulations. If anything is unclear, do not be afraid to ask.

1.1.2 Basic laboratory preparation, kit and consumables

When working in the laboratory, it is necessary to wear a laboratory coat and any other appropriate personal protective equipment that is relevant to the experiments you are carrying out (e.g., gloves and goggles) (check local laboratory guidelines and SDS sheets for any reagents). In addition, you should make sure you know where the closest hand-washing basin is, because you will need to wash your hands before leaving the laboratory. It is a good idea to make sure you know where the eyewash stations are, where any spill-kits are located as well as how to handle the equipment within these kits. This information will normally be covered in any laboratory induction you undergo. Always prepare a risk assessment before planning or starting experiments and discuss this with the laboratory manager before commencing work.

Throughout the course of any laboratory experiment or project, you will need to maintain a clear, coherent, up-to-date record of what you do. This will take the form of a lab book, which will contain all of the protocols you use, results (raw data and subsequent analysis of results), interpretations and conclusions. You can include calculations, diagrams of the experimental setup, negative results, reagent recipes, etc. Your lab book is the most important piece of lab equipment you have – do not lose it. Make sure you begin each new day by writing the date, ensure that you give a title to each new experiment or protocol, and most importantly, keep it up to date. Fill your lab book in as you work—do not make notes on scraps of paper to write up later or hope you will remember what you did, because you will invariably forget, which leads to unnecessary mistakes and repetition of work. Your lab book is not intended to be a work of art, but a record that you can understand and use as a reference for the duration of your project.

When undertaking any microbiology experiments, it is of paramount importance to ensure that all consumables and reagents you use are sterile. Most commercially available kits provide reagents or plastics that are already sterile, so there is no need to worry about these beyond ensuring that you do not contaminate them (see Section 1.5). Ensure that your workspace is free from clutter and that you have enough room to accommodate everything you need. Wipe the bench down with a disinfectant before you begin work and keep a bottle of disinfectant nearby at all times to clean up spills.

For bacteriological work, you might be using a Bunsen burner (check your local laboratory rules), so make sure it is conveniently located in the centre of the workspace and away from anything that might catch fire. If you are using methanol to flame sterilise any equipment, keep the volume small (20–40 mL) and in a glass container or beaker with a lid (or aluminium foil covering it). If you accidently set fire to the methanol, you can cover the container, which will starve it of oxygen and extinguish the fire.

If you are using pipettes for your experiment, keep them within reach of your work and have somewhere easily within your reach to dispose of plastic tips, serological pipettes or glass Pasteur pipettes (check your local lab guidelines for disposal). This is also true for liquid/reagent waste (see Fig. 1.3 for an example of workspace setup).

If you are using media or any other solutions, they should be sterile (see Section 1.5). Try to position these close to the Bunsen burner, and if possible, aliquot them into smaller portions before starting the experiment so that if you do contaminate them, you have not contaminated an entire batch or media/solution.

If you are using a rack to hold tubes or a microtitre plate, position this in front of the Bunsen burner so that it will be in good proximity to the flame, but at this stage, ensure all lids remain in place. If you have reagents that need to be kept cold, place them into a polystyrene box (or similar) containing ice, and position this within easy reach of your work space. For these reagents, remove them from the fridge or freezer and place on ice only when you are ready to begin the experiment.

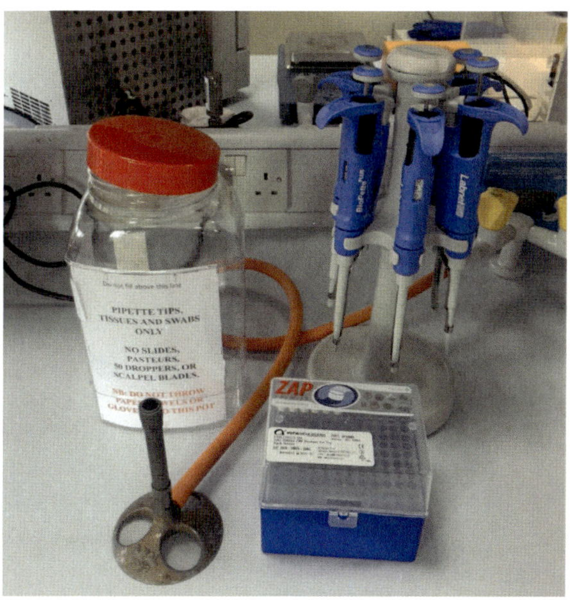

Figure 1.3 Example of workspace setup. Pipettes and tips are within good proximity of the Bunsen burner. Somewhere to dispose of waste is also within easy reach.

Depending on the experiment you have planned and the reagents you are using, you might need to wear gloves and/or eye protection. Check the SDS forms for all reagents you use and your local laboratory guidelines. If you wear gloves and use a Bunsen burner, be careful not to move your hand too close to the lighted Bunsen because the gloves could catch fire. Also, check which of your reagents are flammable, because this will determine how you handle them during your experiment (see Section 1.5). Some aspects of your experiment might require you to handle certain reagents in a fume hood, in which case you will need a way to transport your experiment from your bench to the fume hood. This might require you simply to replace the lids of the tubes or plates, or you might need to use a sealed container such as a Clip and Close box for transport.

1.1.3 Pipettes

Accurate varied volume pipettes (air displacement pipettes) are part of the bacteriological laboratory that you are going to come across often. It is important that you be familiar with your pipettes and how to use them, to ensure that you achieve accurate reproducible results. Using the pipette and selecting the right volume and correct tip to go with it will become second nature to you once you are under way but may cause some initial confusion. If they are used incorrectly, they can cause errors in your work. When using a pipette, one of the first things you should check is the maximum volume the pipette can take up. It sounds simple, but it is a common error for people to use the wrong pipette or to try and force the pipette to above its maximal volume (which can end with a broken pipette). If you look at your pipette, there should be a number or series of numbers on it (Fig. 1.4).

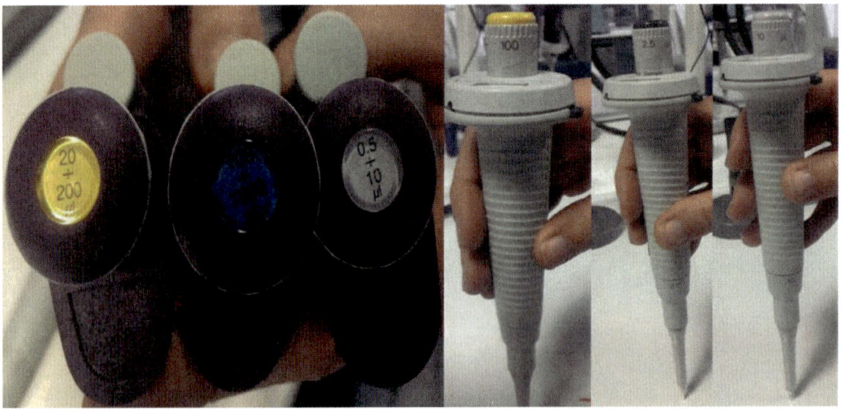

Figure 1.4 Pipettes with maximal volume shown on either the top of the plunger or the side of the plunger. Volumes are given in microlitres.

The largest number is normally the maximal volume the pipette can take. Once you know this, it means you can decide whether the pipette is the one you need for your work. The numbers can be in different places on the pipette, depending on the brand, but they are normally found either on the top of the push button or on the side near the top of the pipette.

Once you know the maximal value of the pipette, you need to locate the window on the pipette where the moveable numbers are displayed, to ensure you know what the maximal volume looks like when it is displayed there (Fig. 1.5). The mechanism for changing the set volume in the pipette will depend on the brand you are using. Often a wheel is on the side or top that moves the numbers or volume, and sometimes a button needs to be pressed while you turn the head of the pipette. Look in the manual for the pipette to make sure you know how this operates before you start work with it.

When working with liquid volumes, you will need to be familiar with how microlitres convert into millilitres, because the volume on the pipettes is given in microlitres and your calculations will often be in millilitres. Figure 1.6 should help you become familiar with the conversions and what they will look like on the pipette.

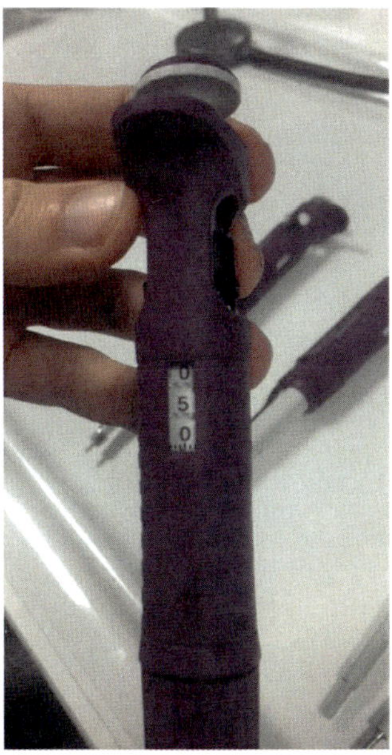

Figure 1.5 Pipettes with working volume window showing. The number in the window should be interpreted in conjunction with the maximal volume of the pipettes. The image shows a 200-μL pipette; therefore, the working volume in window is 50 μL.

Volume µl	Volume ml	Pipette to use	How it will look in pipette window
1000	1	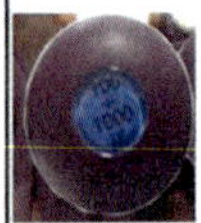	
200	0.2	 or 	 or 
20	0.02	 or 	 or 
2	0.002		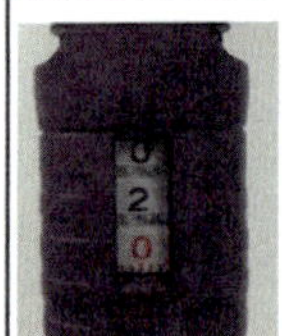

Figure 1.6 Conversion of microlitres to millilitres and how those volumes might look on the appropriate pipette.

A huge range of both manual and electronic pipettes is available as well as single and multi-channel pipettes. The type of work you are doing will dictate which you are most likely to use. Electronic pipettes have a button that controls the takeup and release of liquid (often it can dispense in aliquots). With manual pipettes you are responsible for taking up and dispensing liquid. It is worth familiarising yourself with the pipettes you are going to use. Some have a plunger that goes down one step and back up to pick up liquid, but then needs to be depressed through two steps to expel the liquid properly (you will feel the 'steps' as slight resistance as you depress the plunger). Practice or read the manual to be sure you are doing it correctly before you start work.

There are a few rules that you can follow to make your work with pipettes flow more smoothly. After selecting the volume of pipette you want and ensuring the pipette is set to pick up that volume, you then pick up the tip that fits, making sure that the tip fits with an airtight seal before moving on. Once you have the tip on the pipette, do not put the pipette down on the bench or touch the tip to anything other than the sample with which you are working or the vessel to which you are transferring the liquid. If you touch the bench, your lab coat or anything else with the tip, you will contaminate your work. If you accidently touch something with the pipette tip, eject it into the waste bin and get a new one. When picking up your liquid sample, you only need to put the tip a couple of millimetres into the liquid. If you push it too far into the liquid, you risk picking up liquid on the outside of the tip that could transfer with your sample, giving you an inaccurate volume. You should make sure that tip is a few millimetres into the liquid, though, because if the tip is not in far enough, you will draw air into the tip, which will leaving you with the wrong volume of liquid. When using a manual pipette, it is important to use a slow, smooth action on the plunger so that the liquid you are picking up has time to move up the pipette. Do not put the pipette into a liquid sample without a tip on or you will contaminate and damage the pipette. Once you have dispensed the liquid sample, you will need to remove the tip. You do not need to touch the tip to do this because pipettes normally have an ejector assembly, so you just push the ejector button over the waste bin and the tip will come off. You should always use the ejector button when it is available because you do not want to risk getting bacteria or chemicals on your hands by manually removing the tip. Some larger volume pipettes do not have an ejector button. For these, remove the tip by grasping it at the end closest to the pipette and away from the part that has been in the liquid sample. Pull it off the pipette and dispose of the tip in the disposal bin.

When pipetting very small samples (0.5–10 μL), it can be difficult to see whether you have successfully picked up or deposited the sample. After picking up the liquid sample, look closely at your tip to make sure you have the sample. When dispensing the liquid, it can be useful to tilt the receiving vessel so that you pipet the sample onto the wall of that vessel (Fig. 1.7). This enables you to check visually that you have deposited the sample. With very small volumes, if you do not use the wall of the receiving vessel in this way, sometimes samples can stick to the end of the tip and be lost from the experiment if you throw the tip into the bin. If you are unfamiliar with handling pipettes, you can practice with water or a coloured liquid before working on the samples.

One of the most important things to remember is that once you have liquid in the tip of the pipette, it should remain vertical. If you tip the pipette on its side or upside down, you risk contaminating the inside of the pipette with your liquid or culture, which could contaminate all of your future work.

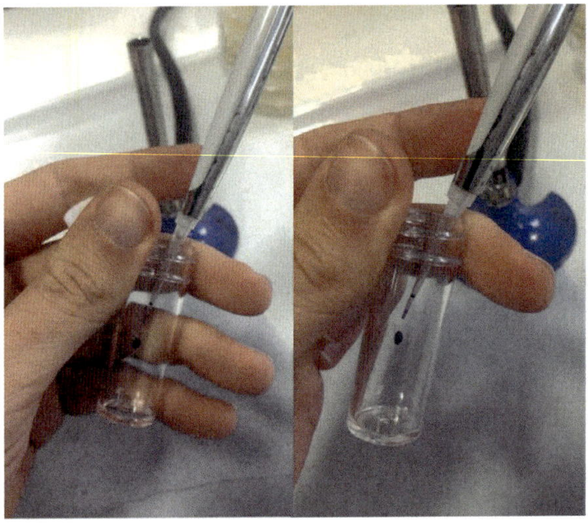

Figure 1.7 Pipette with small volume (2 µL) deposited on the wall of the receiving vessel so it can be confirmed that the small volume of liquid was successfully deposited.

1.2 Aseptic technique

When handling bacterial cultures, it is imperative that you do not contaminate them or any sterile reagents you are handling, and that you do not infect yourself. Aseptic technique is a critical part of successful microbiological work. It is important that you familiarise yourself with it before commencing work, even if this means taking time to practice before you begin your actual experiments.

Any media or solutions you have prepared need to be sterile; mostly these can be sterilised by autoclave at 121°C for 20 min. However, some reagents cannot be autoclaved or need to be autoclaved at a lower temperature, so check this when you are preparing your materials. It is tempting to make up large volumes of reagents, and it is okay to do this, but in practice it is best to aliquot these into smaller volumes before sterilising them. This way, if you inadvertently contaminate a portion of media, you have only 'lost' a small volume and not the entire batch. Make sure that you clearly label the reagents you make, and when you use a portion, write on the bottle the date when you opened it.

Some plasticware is pre-sterilised when you purchase it (for example, Falcon tubes), but some is not. Check before you begin work, and if necessary, sterilise the plasticware you require, again checking the appropriate temperature for autoclaving. Typically, microcentrifuge tubes can be autoclaved, as can plastic

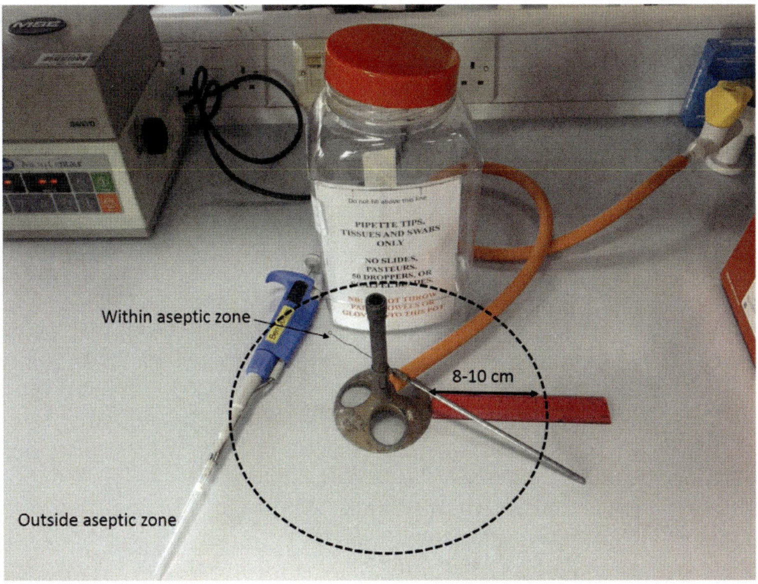

Figure 1.8 The aseptic zone. Once the Bunsen burner is lighted, anything that needs to remain sterile should remain within the aseptic zone (8–10 cm from the Bunsen burner). In this image, the wire loop is within the aseptic zone and therefore is considered to be sterile, whereas the pipette tip is outside the aseptic zone and therefore is no longer considered to be sterile.

pipette tips. For ease, you can fill empty coffee jars with microcentrifuge tubes for autoclaving, and refill tip boxes with non-sterile tips before autoclaving. When preparing any reagent or consumable to be autoclaved, place a small piece of autoclave tape onto the item. This will change colour in the autoclave so you will know what is sterile. It is possible to wrap glass or metal items in aluminium foil before autoclaving. Most items will be wet when they come out of the autoclave, so you will need to dry them in a drying cabinet before use.

If you are preparing agar, remember that it will set if the temperature drops much below 50°C, so make sure that you switch on a water bath (set to 50°C) before autoclaving the agar and transfer it into the water bath as soon as you take it out of the autoclave. All items will be hot when they have been autoclaved, so make sure you wear heatproof gloves when handling freshly autoclaved items; check your local laboratory guidelines with respect to this.

When handling sterile reagents, you will need to practice aseptic technique all the time. This requires you to work within the vicinity of a lighted Bunsen burner. Typically, do not move the item or reagent you are handling more than 8–10 cm away from the Bunsen burner. Any farther than this will be outside the 'aseptic zone' of the Bunsen flame (Fig. 1.8).

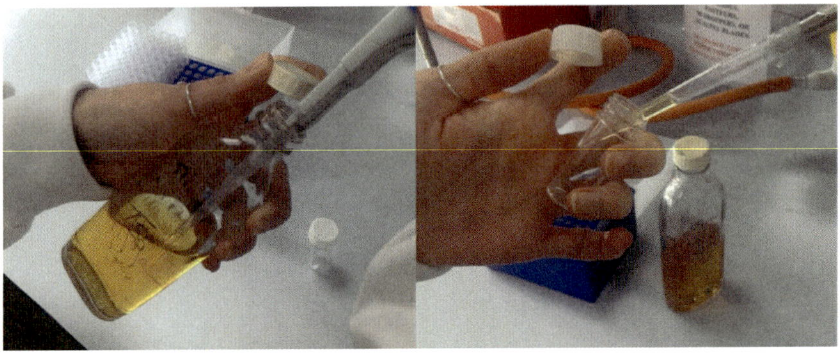

Figure 1.9 Dispensing sterile media from a bottle to a 5-mL sterile bijou. Notice that the lid is held in the same hand as the bottle at all times so that it does not get contaminated, i.e., by being put on the bench. The sterile bijou is close by and within the aseptic zone. The lid has been loosened for ease of removal. Both the bottle and the bijou are angled toward the Bunsen flame (not shown in this image) so that no contaminants fall into the bottle. Notice that once the media have been aspirated from the bottle, the lid is replaced and the bottle is returned to the bench.

When undertaking aseptic technique, the Bunsen flame will need to be set to blue. Do this by opening the air hole on the neck of the Bunsen burner. When you are not undertaking work, either change the flame back to orange or turn the Bunsen burner off so that no one accidently touches the flame. It can be hard to see the blue flame.

When transferring media or solutions from a glass bottle, it is necessary to pass the neck of the bottle through the flame of the Bunsen burner quickly, once away from you and once toward you. Do not flame the neck of any plastic containers; they will melt. Check whether any reagents are flammable. If they are, do not flame the neck of the container. When you are working with media or reagents, it is important to minimise the amount of time the bottle or container is open and to keep it close to the Bunsen flame. Do not remove lids and place the item on the bench. The following description and Fig. 1.9 describe how to transfer 1 mL of media from a 100-mL bottle to a 5-mL bijou.

You will need to take the bottle of media in one hand and a pipette in the other. With the hand that is holding the pipette, remove the lid of the media bottle (loosening the cap in advance is helpful) and hold it. Flame the neck of the bottle of media. Aspirate 1 mL, flame the neck of the media bottle again and replace the cap; you can now put this bottle down. Pick up the 5-mL bijou in your now-empty hand and remove the lid using the hand that still holds the media-filled pipette. Dispense the media into the bijou,

Figure 1.10 Microtitre plate propped up against the base of the Bunsen burner to ensure it remains within the aseptic zone. Once the Bunsen burner is lit, it will be safe to remove the plastic lid of the plate to fill the wells with sterile media.

replace the lid and put the bijou down. Dispose of the used tip into a waste container. The same technique is applicable for transferring media between containers and dispensing media into microtitre plates. For the latter, it is not possible to flame the microtitre plate. Instead, set it down close to the Bunsen flame and keep the lid in place until you are ready to fill the wells. Fig. 1.10 shows how best to position a microtitre plate to ensure it remains sterile while you work.

Pouring agar plates is a vital skill that is done aseptically. Keep sterile agar in a 50°C water bath until you require it. The subsequent description and Fig. 1.11 explain how to pour agar plates.

Prepare the plastic Petri dishes you will fill with agar. Label the plates with the name of the media that will go into them and place them in stacks of five near the Bunsen burner. When ready to pour the agar, remove the bottle from the water bath and dry it. Place it near the Bunsen flame. Position the first stack of plates in front of the Bunsen burner, in the aseptic zone. Take the bottle of media in one hand and remove the lid with the other. Pass the neck of the bottle through the Bunsen flame. With your free

Figure 1.11 Molten agar being poured into sterile plastic Petri dishes. Notice that the stack of dishes remains in the aseptic zone and is filled from the bottom plate upward. This allows all plates to be kept close to the Bunsen flame. The lid of the agar bottle is held in the same hand as the Petri dishes and is not placed on the bench top; this avoids contamination.

hand, lift the stack of Petri dishes so that the lid of the bottom Petri dish is held above the bottom Petri dish. Pour agar into the Petri dish (approximately 20–25 mL) and replace the entire stack. Lift up the lid from the next dish by picking up the remainder of the stack and continue filling the Petri dishes. Pass the neck of the bottle of agar through the flame between each dish. Replace the lid of the bottle of media and put it down when each stack of five Petri dishes is filled. When all Petri dishes are filled, pour any leftover agar into a waste container and rinse out the bottle. Leave the agar plates to set completely; do not move the plates while they are setting (approximately 20 min) or you will end up with lumpy agar plates. If you are using them immediately, dry the plates off by placing them in a 37°C incubator (once set) with the lids ajar (Fig. 1.12) If you are storing them, you can either dry the agar plates before storing or dry them as needed. Agar plates must be stored at 4°C. Refer to the manufacturer guidelines to determine the shelf life of the agar plates.

Sometimes you will need to add supplements to the agar or liquid media. When this is the case, you will need to check whether the supplements can be autoclaved. If they cannot be autoclaved, you will need to filter sterilise them using a syringe and a 0.22-μm syringe filter, as shown in Fig. 1.13. You will need to add supplements to the media using aseptic technique and mix the media thoroughly. Do this by slowly inverting the bottle of

Figure 1.12 Solid agar plates can be dried by placing them in an incubator. By setting the agar plates ajar on their lids, it is possible to avoid inadvertently contaminating them while they dry.

media several times rather than shaking it, to avoid introducing bubbles. It is important to check whether any supplements are heat sensitive and to wait for the agar to be cool enough before adding them. Often, adding supplements will reduce the shelf life of the media. Check with the manufacturer to see how quickly supplements spoil.

This information outlined how to prepare and handle bacteriological media aseptically. However, these rules must be applied to all bacteriological work, even when you are conducting an experiment that uses reagents other than media. When preparing media and reagents for experimental work, consider how much you are likely to need per experiment and make sufficient quantities in advance (when possible), because these can be stored for later use. Regularly check media and reagent stocks to ensure you do not run out during the course of experiments.

Once you decide on the bacteria with which you are going to work, it will be important to prepare and maintain a stock. Stock strains are purchased from a variety of national and international culture stocks such as Public Health England National Collection of Type Culture (https://www.phe-culturecollections.org.uk/collections/nctc.aspx) and the American Type Culture Collection (https://www.lgcstandardsatcc.org/Products/Cells_and_Microorganisms/Bacteria.aspx?geo_country=gb). However, some might be available to you in the laboratory or institution in which you are working. Wherever you source the bacteria, you will need to prepare a working stock and a 'long-term' stock. For the working stock, prepare a streak plate of bacteria (as described subsequently) and subculture it weekly onto fresh media. For long-term projects, it is good practice to prepare a

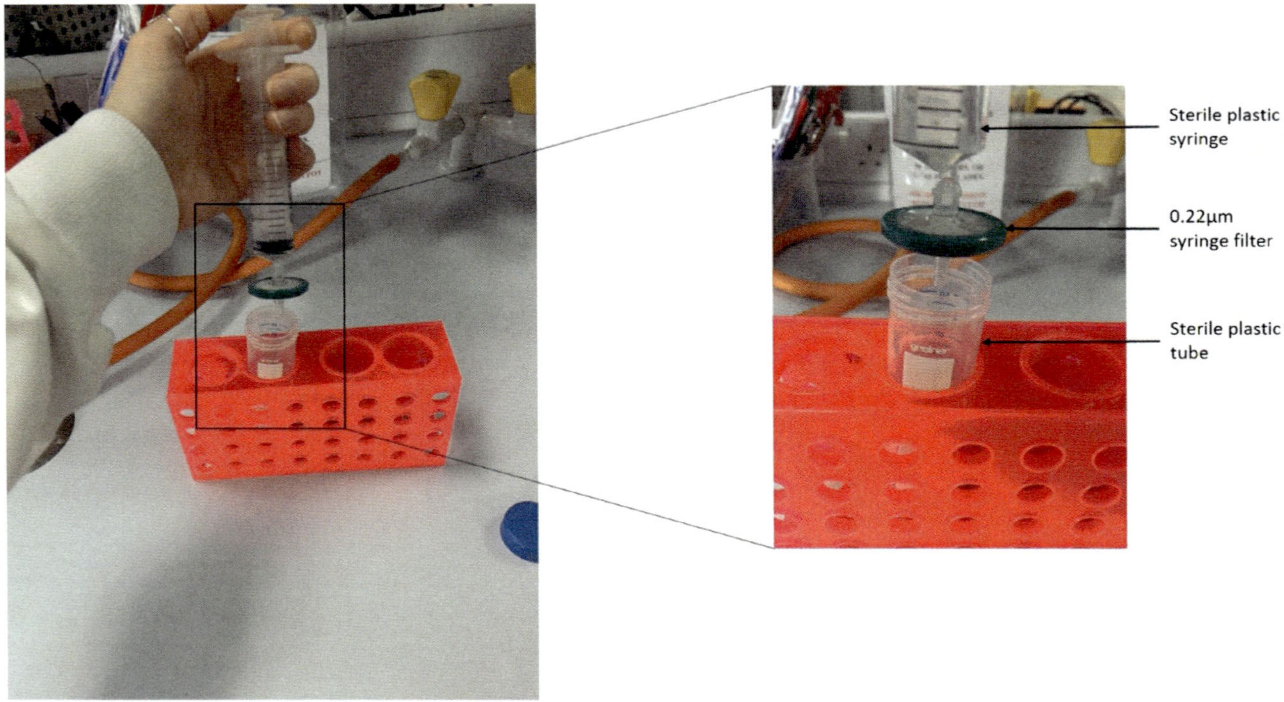

Figure 1.13 To filter sterilise a solution, first draw it into a sterile plastic syringe. Then, attach a 0.22-μm syringe filter to the plastic syringe and dispense the solution into a fresh, sterile container. Do this either in close proximity to a Bunsen flame (within the aseptic zone) or in a laminar flow cabinet.

fresh plate periodically from your original or 'long-term' stock. It will be necessary to discuss the best approach with a supervisor or laboratory manager. Long-term stocks of bacteria can be prepared and stored at −80°C in 50% glycerol. To do this, you will need to prepare a 10-mL broth culture and incubate it at an appropriate temperature for 16 h (check the specific growth requirements of the bacteria you are using and amend accordingly). Centrifuge the culture (speed and time vary depending on the bacteria you are using), discard the supernatant and, working aseptically, resuspend the pellet in 1–1.5 mL of sterile 50% glycerol. Aseptically transfer the suspension to a cryogenic storage tube. If possible, drop the tube into liquid nitrogen to freeze the culture rapidly, and then store the tube at −80°C. Bacterial stocks can also be made using cryopreservation beads designed for long-term storage of microbial cultures. In this case, you would take a colony of bacteria from an agar plate, suspend it in the tube containing cryobeads and liquid, replace the lid and shake the vial for 30 s before removing all liquid from the tube and discarding it. Replace the lid and the beads are ready to go in the freezer. With both of these techniques, prepare more than one stock. Label it with the name of the bacteria and the date prepared. Number the tubes and use them sequentially. Keep an eye on the quantity of the stock culture as you use it and replenish it as necessary.

1.3 Common laboratory units

Irrespective of the type of laboratory project you undertake, you will need to be familiar with units and their appropriate use, to avoid confusion and errors in your work. You will also need to be able to undertake basic calculations when preparing buffers and other reagents for your work. Dry reagents are usually weighed on a balance or accurate balance, typically using grams (g), milligrams (mg) or micrograms (μg). There is 1000 mg in 1 g and 1000 μg in 1 mg. So, 1 mg is 0.001 g and 1 μg is 0.001 mg. You need to remember this and convert among grams, milligrams and micrograms when preparing most standard buffers. When you are preparing solution using dry reagents, the quantity you need to weigh will often relate to the molarity of that component in the reagent. Therefore, you need to be familiar with calculating molarity using mass and molecular weight, which can be done using the formula:

$$\text{Moles in solution} = \frac{\text{Mass of chemical (g)}}{\text{Molecular weight of chemical}}$$

$$\text{Molarity} = \frac{\text{Moles in solution}}{\text{Volume of the solution being made (L)}}$$

The nomenclature for molarity is similar to weight in that there is 1000 mM in 1 M and 1000 μM in 1 mM, and so forth.

Some reagents are prepared using percent solutions rather than molarity. If this is the case, you need to calculate how much dry or liquid reagent you need to use and the total volume of the solution you need to prepare. For these solutions, it is not necessary to know the molecular weight of the compound or reagent you are weighing or measuring. For example, you might need to prepare a 15% solution of sucrose in water. Perhaps you need 250 mL for the experiment you plan to carry out. The amount of sucrose you need to weigh out is 15% of 250 mL, which is 37.5 mg. For this type of calculation, ensure the units, e.g. millilitres and milligrams, are the same. If you were to prepare 2.5 L of 15% sucrose solution, you would need to weigh out 37.5 g. The principle is the same for liquids: if you prepare a 15% solution of glycerol and require 250 mL, you would measure 37.5 mL glycerol and add it to 212.5 mL water (37.5 + 212.5 = 250).

Often when preparing reagents, you will make a stock solution of a particular component and dilute it along with several other components to create a specific buffer. Therefore, it is necessary to be able to calculate how much of a stock solution(s) to add to a given volume to achieve the correct concentration. There are several ways to do this. An important starting point is to decide on or know the final volume of the buffer you are making. You also need to know the 'final' concentrations of each component in the buffer as well as the concentration of the stock ('start') solution. It is important to express the 'start' and 'final' concentrations and volumes in the same units: for example, if you choose to use millimolar for the concentrations, use millilitre for the volume. Table 1.2 shows how to calculate volumes to add to a multicomponent buffer.

Note that although the final volume is 500 mL, the initial volume is prepared as 400 mL (but amounts to a calculation of 500 mL), and the volume is adjusted to 500 mL after the solution has been pH adjusted. This is because adding acid or alkali when the pH of the solution is adjusted will increase the volume.

A simple formula to remember how to do this type of calculation is:

$$\frac{\text{Molarity you want}}{\text{Molarity of stock solution}} = \text{Volume of stock solution to add}$$

Table 1.2 How to prepare a standard 1× solution of phosphate-buffered saline from stock solutions of individual components.

Phosphate-buffered saline (500 mL)

Ingredient	Molarity required	Stock solution	Amount to add
NaCl	137 mM	1 M	68.3 mL
KCl	2.7 mM	100 mM	13.5 mL
Na_2HPO_4	10 mM	100 mM	50 mL
KH_2PO_4	1.8 mM	100 mM	9 mL
H_2O	N/A	N/A	259.3 mL
Adjust pH to 7.4	Then add water to achieve a final volume of 500 mL		

N/A, not applicable.

Remember, if you use moles for the molarity values, the volume of stock solution to add will be in litres; similarly, if you use millimoles for the molarity values, the volume of stock to add will be in millilitres.

By remembering these methods, you should have no problem preparing any type of buffer or reagent for your experiments. Remember that all buffers must be sterilised before use and often will require an adjusted pH. To adjust the pH of a solution, you will require a pH probe and appropriate sticks of acid and alkali. Briefly, you use the pH probe to ascertain the initial pH of the solution. To adjust the pH, you add acid or alkali in a dropwise manner while slowly stirring the solution. You monitor the pH change as you add the acid or alkali. Be patient between each addition and wait for the pH meter reading to become steady before adding more, to avoid adding too much.

In addition to understanding the units needed and preparing reagents, you will regularly encounter a number of other units in the world of bacteriology. These are often used to monitor and/or describe bacterial growth. A commonly used method to measure bacterial growth rapidly in real time is optical density (OD) (see Chapter 2). The OD of a bacterial culture grown in broth is monitored using a spectrophotometer with readings taken at specific time points for a set period. The OD number produced by the spectrophotometer is based on the light scattering of bacterial cells in suspension. The higher the OD number, the more bacterial cells are present in the culture. OD can be specific to different bacteria (because cells may be of different sizes) and also to the wavelength used to take the measurement. For *Escherichia coli*, a wavelength of 620–650 nm is used to measure OD. OD values can be used to prepare graphs to analyse the growth rate of bacteria under different conditions, as described in Chapter 2. Despite this, OD values are not quantitative, in that they do not indicate how many bacteria are present in culture. To determine this, the colony forming unit (CFU) is used. It is possible to devise experiments that equate OD to CFU for specific bacteria under defined growth conditions. CFU is a quantitative measure of the number of bacteria, usually in liquid culture. It requires samples of culture to be taken and grown on agar (see Chapter 2). The number of colonies is counted and the CFUs are calculated accordingly. This method is quantitative and can be applied to any number of experiments that assess microbial growth under a variety of growth conditions. It can also be used in conjunction with OD. As a note of caution, OD must not be confused with absorbance. OD is the amount of attenuation that occurs when light passes through an optical component, taking into consideration both absorption and light scattering, whereas absorbance is a measure only of the former.

Thus, now you should be nearly ready to head into the laboratory and start on your experiments. It is worth considering what resources you will need to carry out your specific experiments while you are there and to make sure you have them close by before starting. To help you, a list is provided of some commonly used resources and their uses (Table 1.3).

Table 1.3 Resources commonly required when working in bacteriology laboratories and corresponding procedures that require them.

Resource	Procedure
Personal protective equipment: Lab coat/goggles/gloves	Protection of user, prevents contamination of clothing/splashing from chemicals
Lab book	Record of all laboratory activity
Permanent marker pen	For labelling cultures/bottles/autoclave type
Diamond marking pencil	For labelling glass slides
Glass slides plus coverslips	Microscopy/staining
Petri dishes	For agar plates
Bunsen burner	To assist aseptic work
Inoculating loop (wire or plastic)	Streak plates, inoculating cultures
Spreader	Lawn plates
Universal container	Broth cultures
Glass bottles (500 mL)	Sterilising media (agar and broth)
Staining rack (gram stain reagents)	Microscopy
Disinfectant	Wiping down bench, clearing up spills
Autoclave tape	To indicate sterilisation
Pipettes	Measured transfer of liquid
Pasteur pipettes/teat	Drop transfer of liquid

Bigger pieces of equipment such as autoclaves, incubators, fridges, freezers and water baths are likely to be shared within the laboratory and are not kept on your bench. It is worth checking whether these are available and whether there is a booking system for any of them. In busy laboratories, some equipment will be in heavy use. Once you have assessed the risks, filled in the paperwork and ensured you have all the equipment you need, you will be ready to enter the lab.

1.4 Top tips for working safely in the lab

- Treat all microorganisms as potential pathogens; work aseptically.
- Dress appropriately, wearing a laboratory coat, gloves and safety goggles as needed.
- Tie long hair back and do not wear open-toed shoes.
- Sterilise all equipment before you start your experiment.
- Disinfect your bench before and after you use it.
- Make sure you know where the safety/spill equipment is and have your COSHH assessment nearby.
- Label all of your work clearly (make sure you have a permanent marker), including any hazard warnings that are needed.
- Keep an accurate record of everything you do in your laboratory notebook.
- Do not eat or drink in the laboratory and avoid hand to face movements.
- Dispose of waste in the correct manner; all bacterial waste will need to be autoclaved.
- Remove your laboratory coat and wash your hands before leaving the laboratory.

1.5 Notes page

Record observations and notes here.

CHAPTER TWO

Bacterial growth in solid and liquid media

Contents

2.1 Introduction

The ability to grow and culture different species of bacteria is an essential skill you will need in microbiology. Bacteria are ubiquitous in a vast range of conditions including in soil, water and food, as well as on and inside the human body. As you can imagine, each of these types of bacteria has its own growth requirements and can be cultured successfully only if you provide the correct balance of nutrients in conjunction with appropriate growth conditions, such as optimal temperature and oxygen levels. Once growing, it is essential that you separate the bacteria into individual colonies on agar. This allows you to observe individual characteristics of bacteria and can help identify unknown species within different samples. There is a huge range of media from which to choose when growing bacteria. Many of these contain specific substrates incorporated to aid in identifying bacteria and which reveal the individual characteristics of bacteria. Even if you already know which bacteria you are working with, you still need to be capable of culturing single colonies, because this allows you to confirm that you have not contaminated your work and provides pure single colonies that can then be used to inoculate more complex experiments. This chapter will help you understand the techniques and conditions you need to achieve and maintain pure cultures of bacteria. It also describes the different ways in which you can measure the growth characteristics of your bacteria of interest.

Bacteriology Methods for the Study of Infectious Diseases
ISBN 978-0-12-815222-5
https://doi.org/10.1016/B978-0-12-815222-5.00002-X

2.2 Media, culture conditions and nutritional requirements

To cultivate bacteria of interest successfully within the laboratory, you will need to provide growth media. These normally come in either agar or broth form (sometimes semisolid agar can be used), which can have highly varied compositions depending on what bacteria they are designed to support. The nutrient requirements of different bacteria can vary widely and the media and supplements you choose can have a profound effect on bacterial growth.

Generally, work with media in the laboratory falls into two broad categories: chemically defined media (also known as synthetic media) and undefined media (also known as complex media). Examples are given subsequently of these media and their main uses.

Chemically defined media: As the name suggests, this type of media is created using a specific and known set of ingredients so that the exact composition of the final product is known. They are often used to grow bacteria that have specific growth requirements. They include media such as M9 Minimal Salts Broth and glucose salts broth.

Complex media: This type of media contains one or more components that are not completely characterised. They can have a varied composition and tend to be rich media that provide a wide range of bacteria with the amino acids, sugars and other nutrients they need to grow. These undefined components can come from plant or animals sources and may include yeast or beef extract. Examples include nutrient agar/broth, tryptic soy agar/broth, Luria-Bertani agar/broth and brain heart infusion agar/broth.

These media are designed to support the growth of a wide range of bacteria. If you select the right one, you will be able to use it to maintain your cultures daily by producing streak plates, which allow the isolation of single colonies. It will also help you to identify unknown bacteria from a sample, such as from a skin swab. For bacteria to grow properly either in the broth or on the agar you have provided for it, you will also need to control some environmental factors. Therefore, you need to know the optimal range of temperature, pH and oxygen level required to grow the bacteria. If you are trying to grow anaerobes, you will need to work within an anaerobic cabinet; if your laboratory does not have one, you could work with anaerobic gas jars, which, used in conjunction with special sachets, create an anaerobic environment inside the jar.

2.2.1 Isolation

When working in bacteriology, it is essential to be able to culture bacteria to obtain single colonies on agar. You will need single colonies if you want to be able to characterise the bacteria with which you are working and to be confident that there is no contamination of the culture. To achieve this, it is normal to spread and dilute the bacteria. This is often done using the streak plate method.

To perform a streak plate, you will need to continue working aseptically and have the culture from which you want to work (this could be another agar plate, broth culture or −80 stock culture), a sterile agar plate, a Bunsen burner and an inoculating loop (either a sterile plastic one that can be disposed of after each use or a metal one that can be sterilised in a blue flame between streaks).

Before starting, you should ensure that the sterile agar plate is labelled so that you can identify it after incubation. It is normal to include the media type, your name, the date and the name of the organism you are culturing. Do not forget to write around the edge of the plate and on the bottom of the plate, not the lid. This will allow you to see the agar and culture through the bottom of the plate so that you can identify each plate even if the lids are dropped or mixed up. When you are ready to make a streak plate, you need to light the Bunsen burner and turn it to the blue flame. It also makes things easier if you invert the agar plate so that the lid faces down on the bench. Then, as you pick the plate up to streak the bacteria onto it, you do not also have to hold on to the lid; you can replace the plate back onto the lid each time you need to flame the loop between streaks.

If you use a metal inoculating loop, you need to place the loop end of the inoculating loop into the flame until it glows red; this sterilises the loop. If you look closely at the flame, you should see a smaller blue cone of flame within the larger flame. The tip of this inner cone is the hottest part of the flame and the best place to flame the loop.

Once the flame glows red, you can remove the loop and wait a few seconds for it to cool. This cooling step is important: if you were to take the loop from the flame and immediately touch it to the bacterial sample, the heat of the loop might cause aerosols, which could be dangerous. You would also kill the bacteria on the loop you were trying to culture.

Once the loop has cooled, you can use it to pick up the bacterial culture. If you are taking bacteria from another agar plate or a broth culture, you do this simply by touching the loop to a single colony on a plate or taking a loopful of liquid from a broth culture (remember to work aseptically, flame

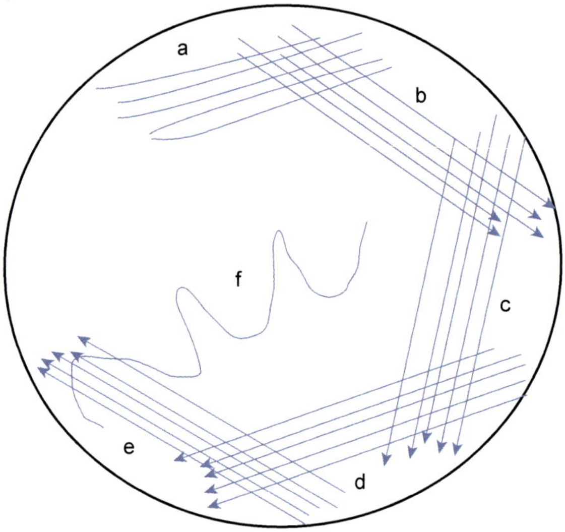

Figure 2.1 Diagram of how to prepare a streak plate to obtain pure cultures of bacteria. Between each set of streaks, the loop should be sterilised if it is metal or disposed of if you are using sterile plastic loops.

the neck of any glass bottles you are using and do not put lids down on the bench). Then, transfer this to the sterile agar plate and gently move the loop over a small area of the agar plate (a in Fig. 2.1). You have now inoculated your agar plate.

At this point, you need to flame the loop again until it is red. Wait for it to cool and then use the loop to streak the inoculum in one direction, starting at the initial inoculum site and moving away from it along one edge of the agar plate (b in Fig. 2.1). At this point, you need to flame the loop again. Wait for it to cool, and this time streak the bacteria from the end of the first set of streaks you made, again in one direction away from the first streak (c in Fig. 2.1). You continue flaming you loop and making new streaks until you have moved most of the way around the agar plate (d and e in Fig. 2.1). The last time, you can flame the loop and cool it, and then make a single streak from the last set of streaks, up the middle of the agar plate, making sure not to touch any other streaks or the initial inoculum site.

As you progress through the streaks, you will eventually get one single bacterial cell on the surface if the agar, which will grow into a singly colony when incubated (Fig. 2.2). The time and temperature at which you incubate the bacteria will depend on the bacteria with which you are working, so you should check the literature for the common growth requirements of

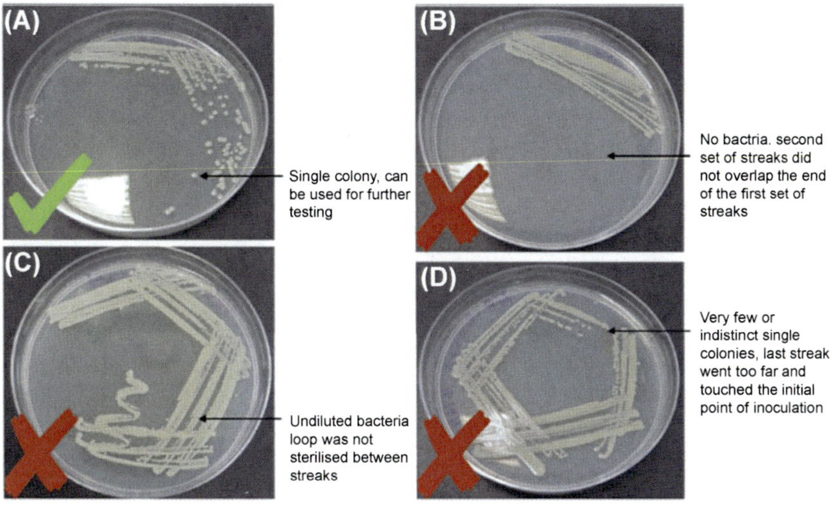

Figure 2.2 Agar plates showing (A) single colonies of bacteria, providing a pure bacterial culture to work with; (B) an agar plate on which the loop was not flamed between streaks, leading to an undiluted culture with no single colonies; (C) an agar plate on which the second set of streaks did not cross over the first set of streaks, leading to no bacteria being spread beyond the first streak and no single colonies; and (D) an agar plate on which the last streak touched the original inoculation, leading to no or few distinct single colonies. The *green check mark* indicates how the plate should look. *Red x's* indicate plates that have mistakes.

your particular bacteria. Once you have single colonies, they can then be subcultured onto other media, investigated for their characteristics or used to make frozen stocks.

If using plastic inoculating loops, dispose of the original loop between each set of streaks instead of flaming, and start each new set of streaks with a fresh sterile plastic loop. It is important either to flame the metal loop or dispose of the plastic one between streaks, because the point of this technique is to dilute the concentration of bacteria gradually as you are move along the plate. If you do not flame or dispose of loops, you will end up with too many bacteria on the plate and will not isolate single colonies.

If you are making a streak plate from frozen culture, the process is somewhat different. If employing glycerol stocks, use the loop to pick up a small amount of frozen culture and then continue with the procedure as outlined previously. If making a streak plate from frozen beads, you will need to remove one bead from the frozen vial and place it onto the agar plate. This sounds simple, but sometimes it can be difficult to remove the bead from the vial because the beads are often small enough to slip through an inoculating loop. If you are struggling to get a bead out of the vial, it is sometimes useful to use either sterile forceps or a sterile 1-mL pipette tip, which can be inserted into the hole in the middle of the frozen bead. This will allow you to move the bead from the vial to the agar plate. Try to complete this quickly because you do not want the stocks to thaw out on the bench. You could place the samples in an ice box while you are working on them, to keep them cool. Once you have removed the bead from the vial and onto the agar plate, put the stock back into the freezer. If you repeatedly thaw and refreeze the vial, it can damage the bacteria you are trying to store. Next, you will need to move or roll the bead on the agar plate so that you have a discreet area of inoculation. Use a sterile inoculating loop to streak from that area and continue as outlined earlier. You can leave the bead on the plate during incubation or remove it by tapping the plate above the waste disposal container. You may find that bacteria from the freezer take a bit longer to recover than those taken from fresh cultures. If so, you can incubate the plate longer until there is stronger growth.

When first streaking a plate, people encounter other common problems such as damaging the agar with the loop and getting too much or not enough growth. With practice, you can avoid these problems. Nevertheless, here are some tips to help you create a successful streak plate:

1. Make sure the agar is fully set before you start work. If it is still warm or not fully set, you are likely to make holes in it with the loop. It is also best to use the bottom edge of the loop on a wide angle. If you drag the loop along the agar on its side, you might slice into the agar, damaging the surface.
2. When flaming the loop, you will have put the agar plate back on the bench with the lid on. Sometimes it can be hard to remember where you finished the last set of streaks. If you go over the same streaks, you will end up with growth in one place and no growth in another. Try tilting the plate in the light; this can often help you see where the last streak was.
3. If you do not move the loop through the end of the previous streaks after sterilising it, you will end up with no growth on those areas of the plate. You can minimise the chances of losing your place by lining up the initial inoculum with the writing on the bottom of the plate when you start the process. This gives you a reference point. You can also try tilting the plate in the light, and sometimes you can see where you have run the loop along the surface of the agar. Make sure you retain aseptic technique when doing this.

4. Make sure you flame the loop between streaks. If you forget to sterilise the loop, you will not be diluting the bacteria and will end up with strong growth on all streaks and will not get single colonies.

2.3 Differential and selective media

You might find that when conducting your experiments, you need something to help you separate bacteria from mixed populations into separate pure cultures, to identify or characterise bacteria by their biochemical capabilities or to aid you when you want to enumerate bacteria taken from a mixed sample. To help with this, a range of special media were designed that can differentiate, select, or enrich populations of bacteria. These media, like the ones discussed previously, contain all nutrients needed to grow bacteria, but they will also have added components that allow you to isolate or identify specific bacteria or groups of them.

Selective media can be useful, especially if you are trying to isolate one group of bacteria from another. This is normally achieved when the media contain a substance that will support the growth of one type of bacteria but inhibits the growth of other types. You might use this if, for example, you were trying to isolate all staphylococcal species from a skin swab but do not want gram-negative bacteria to grow.

Differential media, in contrast, allow more than one type of bacteria to grow but help you to distinguish between two or more groups of bacteria based on their biochemical capabilities, usually through a colour change in the media of bacterial colonies. This can be useful if you are trying to identify an unknown isolate. Often, media are both selective and differential.

For example, mannitol salt agar contains high levels of sodium chloride (7.5%), mannitol and phenol red (a pH indicator). The selective part of this agar is the high level of salt, which will allow the growth of staphylococci but will inhibit most other bacteria from growing. The differential part of this medium is the phenol red, which will change colour depending on the pH of the medium, which is altered in different ways by different bacteria. If you wanted to check whether you had *Staphylococcus aureus* or *Staphylococcus epidermidis* within that sample, you could use this medium to answer the question. The medium would support the growth of both staphylococcal species, but the *S. aureus* will metabolise the mannitol, producing acid, which changes the colour of the phenol red in the agar to yellow. In contrast, *S. epidermidis* cannot use mannitol, so it does not produce acid, leaving the phenol red unchanged (Fig. 2.3).

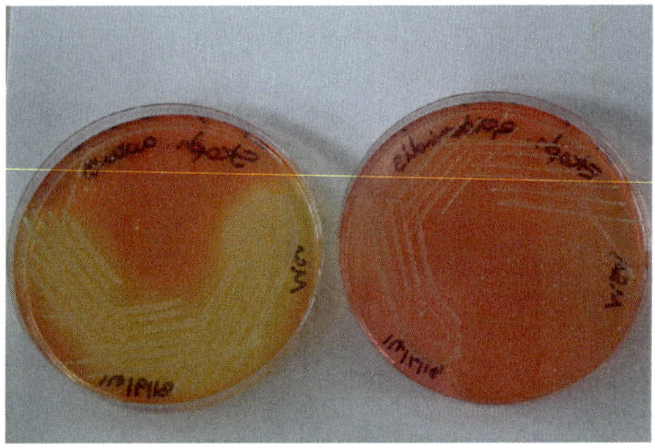

Figure 2.3 Mannitol salt agar plates incubated with *Staphylococcus aureus* and *Staphylococcus epidermidis*. The plate on the left shows the typical yellow colouration produced after incubation with *S. aureus* and the plate on the right shows the pink-red colour seen after incubation with *S. epidermidis*.

MacConkey agar is another example of a selective and differential medium that is commonly used. Selective parts of the agar are the bile salts and crystal violet, which allow the growth of enteric organisms but inhibit the growth of gram-positive bacteria. The differential part of the agar is lactose coupled with a pH indicator called neutral red. Bacteria that can strongly metabolise lactose such as *E. coli* will turn pink-red and produce enough acid to precipitate the bile salts in the agar. This causes the media surrounding the bacterial colonies to turn pink. Bacteria that can ferment lactose, but not as strongly, will cause a pH drop in the agar, turning the bacterial colonies pink-red, but will not cause the agar surrounding the bacteria to change colour. Bacteria that cannot metabolise lactose, such as *Salmonella* spp or *Pseudomonas aeruginosa*, will still be able to grow on the agar and will use the peptone available in the agar. This will lead to an increase in pH, giving the bacteria a white or colourless appearance (Fig. 2.4). If you are trying to isolate and differentiate between lactose fermenting and non-lactose fermenting enteric bacteria, this would be an agar worth considering in your work.

Eosin-methylene blue (EMB) can also be used to select for gram-negative bacteria (often faecal coliforms) because it has selective components consisting of two dyes (eosin and methylene blue) which suppress the growth of gram positive bacteria while allowing gram-negative organisms to grow. In this case, bacteria that can ferment lactose/sucrose depending on the composition of the EMB agar you use (the differential part of the agar) and create a low pH will grow as colonies with a dark purple centre. *E. coli*

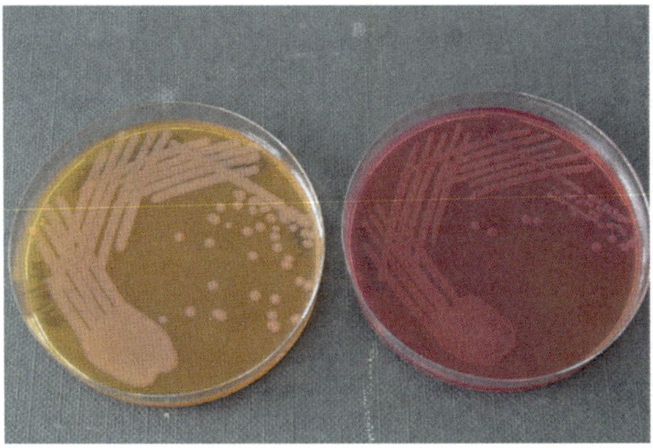

Figure 2.4 MacConkey agar plates incubated with *Pseudomonas aeruginosa* and *Escherichia coli*. The plate on the left shows typical growth on the plate after incubation with non-lactose fermenting *P. aeruginosa* and the plate on the right shows the typical pink colonies and agar caused by incubation with lactose-fermenting *E. coli*.

Figure 2.5 An eosin-methylene blue agar plate after incubation with *Escherichia coli*. Note the metallic sheen seen around the edges of the bacterial growth.

will often grow as colonies with a distinctive metallic sheen and dark purple centre owing to the way it rapidly ferments lactose within the media, creating a strongly acidic environment (Fig. 2.5). The less powerful fermenters can have a pink mucoid appearance on this agar. You may also sometimes find colourless colonies on this agar. These are organisms that do not ferment lactose and are normally not faecal coliforms.

Lots of chromogenic agars are available on the market that will differentiate among bacteria by using chromogenic substrates that are activated when those substrates are cleaved by specific enzymes, so it is worth looking around if you are doing experiments that require isolating and growing specific bacteria. These media will come with instructions that help you understand what the different colours of colony mean. These chromogenic agars can be purchased from companies such as CHROMagar (http://www.chromagar.com/), Sigma-Aldrich (https://www.sigmaaldrich.com/united-kingdom.html) and Oxoid (http://www.oxoid.com/UK/blue/search/results.asp).

Media have also been designed to enrich the growth of bacteria by providing a rich source of nutrients, which might be helpful if you are trying to culture a mixed sample of bacteria or a single type that is fastidious, such as *Haemophilus influenzae*, which will not grow on non-enriched media. Commonly used enriched media include, among others, both blood and chocolate agar (often horse, sheep or rabbit blood). These contain nutrients to support many types of bacteria and blood and heat-treated (lysed) blood, respectively. You can buy pre-poured agar plates that already contain the blood, or you can buy the blood and add it to the base media yourself. With both the premade plates and the blood, it is important to note the expiration date and storage conditions.

With all of the media discussed earlier, as with anything you use in the laboratory, it is a good idea to read through the manufacturer's guidelines regarding how to reconstitute the media you buy. If you buy it in powdered form and make it up yourself rather than buying premade plates or broth, it is also important to check how to store not only the original product (often at room temperature) but also the plates and broths once they are made up. If you purchase premade plates, check the storage conditions and shelf life for those products (often fridge storage).

2.4 Estimating cell number

As you move on with experimentation, it is almost certain that you will need to enumerate bacteria at some point. In general bacteriology, enumeration is frequently used and is particularly important when investigating the number of bacteria found in water or food samples. It is also commonly employed in other strands of research for many purposes from standardising the starting inoculum to determining how many bacteria in a population have survived a novel antimicrobial treatment. The various ways to enumerate will be covered here, starting with one used as a standard in bacteriology: measurement of the optical density (OD) by a spectrophotometer.

As cells grow and divide, the turbidity of a bacterial culture increases. This turbidity can be used to estimate trends in the growth of the culture. When turbidity is dense enough to be detected by eye, this phenomenon is caused by light scattering off cells present in the media. OD measurement detects the amount of light lost owing to scattering and absorption at the wavelength chosen. You cannot tell by eye how many cells are in the population, but you can use a spectrophotometer to estimate the cell number or to follow the growth pattern of the culture over time.

Spectrophotometry is often used in research, especially before inoculating an experiment, to ensure that every experiment has the same number of cells inoculated into it at the beginning. This standardisation of the starting culture allows you to minimise the effect of the starting inoculum on the results. Before starting an experiment that requires a spectrophotometer, make sure your laboratory has one and that you have cuvettes that fit the spectrophotometer that is in stock. If your laboratory does not already have cuvettes in stock, you can get both macro- and microcuvettes (which can be reusable or disposable). If you need to purchase some for your laboratory before starting the experiment, they can be purchased from various suppliers, including Merck (https://www.sigmaaldrich.com/united-kingdom.html), Fisher Scientific (https://www.fishersci.co.uk/gb/en/home.html) and VWR (https://uk.vwr.com/store/;jsessionid=VRhV16QI9jHOBACYAOX96VVH.estore5b).

Once you have checked that you have the right equipment, you can use the spectrophotometer to set up lots of experiments, including basic growth curves and antimicrobial susceptibility assays, or to establish which in media your bacteria grows best. For example, if using the spectrophotometer to normalise cell numbers at the beginning of an experiment, you might follow a procedure like the one described next.

Working aseptically, take a loopful of frozen stock (e.g., *S. aureus)* and inoculate 10 mL nutrient broth; incubate it overnight at 37°C. The following day, from your now turbid culture, take 1 mL of the overnight culture and inoculate a fresh 10-mL nutrient broth and incubate it for a couple of hours at 37°C so that the cells reach log phase growth before you start working with them. At this point, you will need to use the spectrophotometer. Turn on the spectrophotometer at the wall and make sure it is switched on at the machine if there is another button there.

Remove your culture from the incubator, pipette a portion of the cells into the cuvette, make sure you know what volume your cuvette holds and ensure when pipetting that you do not over fill the cuvette. Do not forget to keep working aseptically and place the lid back on the cells you are not putting into the cuvette.

It is important when handling the cuvette that you do not accidently get material on the part of the cuvette through which light will pass.

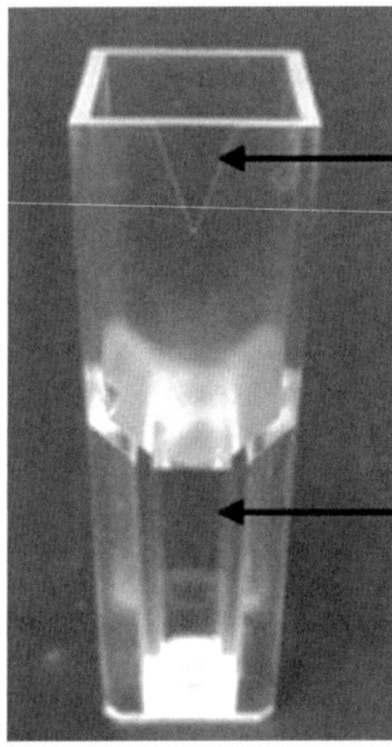

Figure 2.6 Microcuvette shown from the front. Because light passes through the cuvette in this orientation, you should try to keep this area as clean as possible so no dirt will alter the results. Try to hold the cuvette by the sides.

Most cuvettes have a mark on the top that indicates which way round in the spectrophotometer it will go. This means you can hold the cuvette by the frosted sides to avoid getting fingerprints or dirt on the front and back of the cuvette where the light will pass through (Fig. 2.6). You can also wear gloves to prevent fingerprints from transferring onto the cuvette.

In addition to a cuvette containing the cells, you will need a second cuvette that contains a control (blank) liquid. In this example, it would be nutrient broth with no cells, but you should use whatever media you are growing your cells in. Once both cuvettes have the correct volume of liquid, you can measure the turbidity of your cells.

You will need to select the wavelength of light that you are using on the spectrophotometer. Some spectrophotometers read only at set wavelengths, so check what these are before you use the machine. Some allow you to pick your own wavelengths. If you are working with bacteria, 600 nm is commonly used and could be taken as your starting point if this technique has not been employed before in your laboratory. Other wavelengths are used fairly commonly with unpigmented bacteria, such as 480, 540 and 660 nm, so it is worth checking whether your machine is set to one of these or whether the bacteria with which you are working produce pigment that might require a specific wavelength. Once you have selected the wavelength, you should insert the control (blank) cuvette into the machine and

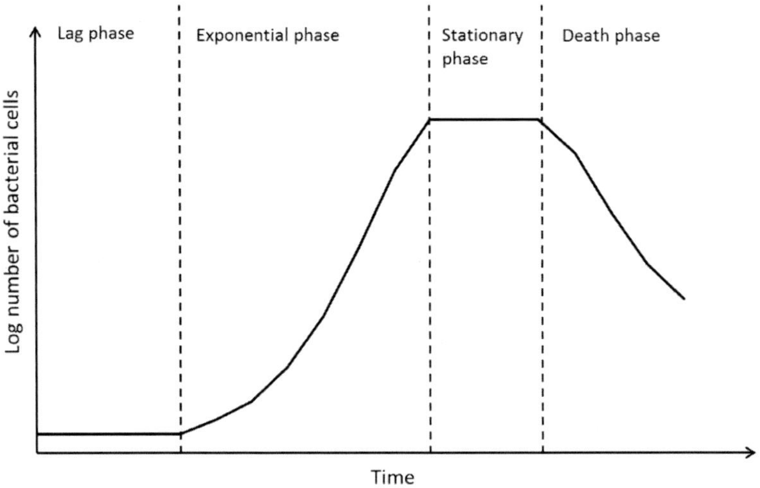

Figure 2.7 Example of a growth curve demonstrating the four main phases of growth you are likely to see during a growth experiment: lag phase, exponential phase, stationary phase and death phase.

make sure the mark on the top of the cuvette is in the correct orientation so light can pass through the clear part of the cuvette to the detector on the far side. Normally, you close the lid of the spectrophotometer and then select zero on the machine.

Wait until the machine has given you a zero reading. Then, open the lid and remove the first cuvette. You can then place the cuvette containing the bacterial culture into the machine, making sure the cuvette is in the correct orientation, and close the lid. With most spectrophotometers, the OD reading should display on the machine where the zero reading was previously. Record that number in your laboratory notebook and do not forget to remove the cuvette from the machine and throw it away once you have the reading.

If you are testing lots of samples, it is a good idea to re-blank the machine between each sample to be sure you obtain accurate readings. Once you have the OD, there are a couple of ways in which you can use it. If you are doing something like a growth curve, you take the reading at each time point and plot it on a graph. Growth curves usually follow a distinctive patterns with four phases: lag, exponential (growth), stationary and death (Fig. 2.7).

If you are using the OD reading to standardise the inoculum to start another experiment, you might first have to take extra steps. For example, if you wanted a starting inoculum of 0.08–0.1 OD, you would measure the OD of your culture, as described previously. If the OD is too high, you might need to dilute the original cells and then recheck the OD, or if the OD is on the low side, you might need to let the culture incubate for longer before checking the OD again. If you dilute the cells, you need to use the same media in which you are growing them; once they have been diluted, check the OD. This way, even if you need to carry out the experiment again months later, you can make sure that the starting inoculum is similar.

You might also use a spectrophotometer in conjunction with microtitre plates. Machines that work with these plates are designed to read multiple wells at the same time and can be used to take end point readings of cell growth in much the same way as the cuvette spectrophotometer (so you could read at 0 and 20 h). Some can be used to hold a plate for a period of time (so you could add the plate to the machines and set it to read every 30 min for 20 h). This can be useful if you want to look at growth over time. In this case, inoculate the plate with the bacterial culture (checking the initial inoculum using the spectrophotometer, as discussed earlier). Then, insert the microtitre plate into the spectrophotometer, It is a good idea to take a reading of the plate at this point (a 0-h reading) because this can be subtracted from any readings you take later, which allows variations in initial turbidity or the colour of the broth to be removed from the growth curve.

These microplate spectrophotometers often have heating and shaking capabilities, so you can tailor the incubation to your needs. Some spectrophotometers will also allow you to select how often and where in the well you take your readings from. They can be useful for generating growth data, especially if you are looking at the effect of a compound on the bacteria, or at how different mutations have affected growth. For this, you can set up the plate in various ways. For example, Row 1 could be your negative control (growth media, no bacteria), Row 2 could have an inhibitor, Row 3 could have a supplement and so on. The last row could be the positive control (growth media and bacteria, no inhibitor) (Fig. 2.8).

Once the plate is set up, put the spectrophotometer in place and take OD readings at specified time points to show how these different conditions affect growth (Fig. 2.9).

Sometimes the spectrophotometer can be used to look at end point readings of stained bacteria. You should check which wavelength is best for that stain. For instance, if you were measuring biofilm growth (details are provided in Chapter 6) using crystal violet, the wavelength is set to 570 nm. Check the literature regarding manufacturers' recommendations for the most appropriate wavelength before you set up the experiment.

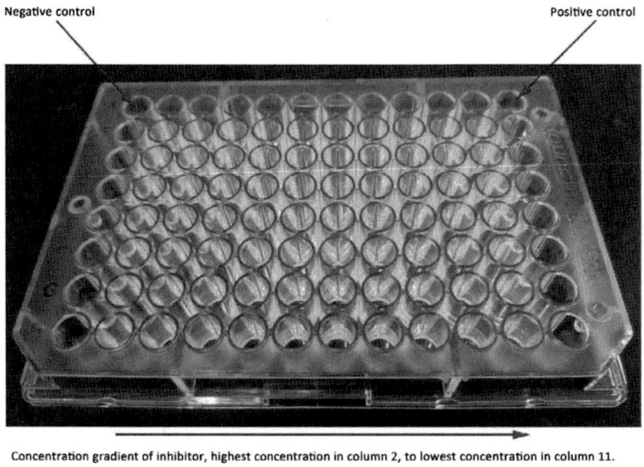

Figure 2.8 A 96-well microtitre plate with basic layout provided. Rows A–H can hold replicates of the sample whereas Columns 1–12 contain both negative and positive controls as well as a different test condition in each column from 2 to 11.

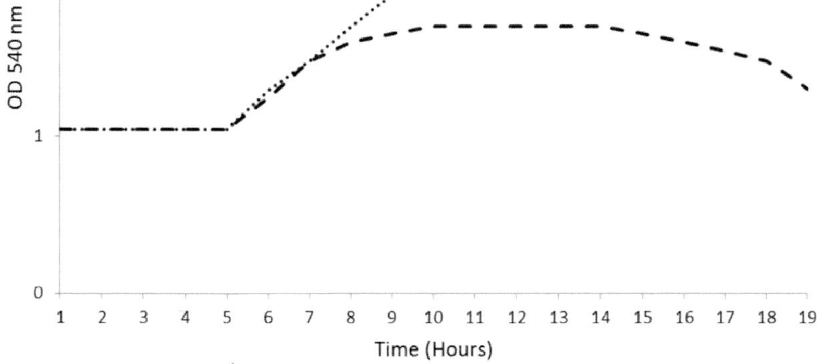

Figure 2.9 Example of what a growth curve might look like when growing bacterial cells under control conditions with just growth media (*dotted line*) and growing bacterial cells when incubated with both growth media and an inhibitor (*dashed line*). Using the spectrophotometer to read the plate at various time points, you can see when the inhibitor has an impact on cell growth. *OD*, optical density.

In addition to using a spectrophotometer, bacterial numbers can be estimated by eye, by using a McFarland standard for comparison with cell growth. McFarland standards are solutions made up of barium chloride and sulfuric acid. You can buy McFarland standards at different levels of turbidity that approximate different numbers of bacteria. To test the culture, hold the culture tubes up alongside the McFarland standard against a white background with black lines to check turbidity. The culture should be as turbid as the McFarland standard to have the number of bacteria approximated by the McFarland standard in it.

The 0.5 McFarland standard is commonly employed; it has approximately the same turbidity as $1–2 \times 10^8$ colony forming units (cfu)/mL. Guidelines provided by the European Committee on Antimicrobial Susceptibility Testing (http://www.eucast.org/) are a good example of how the McFarland standard is used in bacteriology. Within these guidelines, you will see that the 0.5 McFarland standard is used as a guide for how turbid the culture should be before lawn plates are prepared for disc diffusion antibiotic susceptibility testing. (For details on antibiotic susceptibility testing, see Chapter 4.) As suggested in those guidelines, it is probably better to use a spectrophotometer to adjust the bacterial suspension because it can be difficult to judge it by eye and you want the experiments to be as reproducible and accurate as possible. If you want to check how accurate the enumeration of the culture is and do not have access to a spectrophotometer, you can plate out some of the culture as a total viable cell count (see direct enumeration, discussed subsequently) just before or after setting up the experiment. This way, you can incubate the cell count plate overnight and check the cfu per millilitre of the sample the next day to confirm that you started with the right number of cells.

You will probably use both the spectrophotometer and the microplate spectrophotometer a lot when conducting experiments, because they provide quick and easy ways to check that you are using a reproducible number of cells in your experiments. However, the two methods mentioned earlier do not provide a direct count of bacterial cell numbers and only estimate the numbers of cells in the culture. As with most methods, they have limitations. You may encounter some issues when using them, depending on the type of bacteria and the experiment you are performing. When looking at turbidity, either by eye or with a spectrophotometer, you cannot tell whether the turbidity you can see is being generated by live or dead cells, which can lead to an overestimation of numbers if there are many dead cells in the culture. The spectrophotometer readings could also be misleading if the cells have produced extracellular material that can also scatter light and may lead to an overestimation of cell numbers. Some bacteria clump; this can lead to odd readings when using the spectrophotometer. It is worth looking at the tubes or plates by eye before or after you use the spectrophotometer and making a note in your laboratory notebook of any that you think you can see clumping, because they can result in very low or high OD readings. It is also worth checking to see whether you can see a lot of cells stuck to the side of the tube or plate you are using, because bacteria can form biofilms, which might alter the reading you obtain when looking exclusively at OD readings.

2.5 Direct enumeration (viable cells)

There will be times in your research when you will need a direct count of bacterial cells. This can be achieved through a number of methods. For example, if you want to plot a growth curve but you want a count of viable cells in the culture rather than an estimation from the spectrophotometer, you can use a viable cell count method. This technique is based on the Miles and Misra method first described in 1938 and uses a serial dilution technique. The methods allows you to dilute the bacteria, enabling you to obtain a viable count. This is important because bacterial growth is rapid, and if you are counting bacteria from an overnight incubation or even at multiple points for a growth curve, soon there will be too many bacteria to count accurately if they are plated undiluted onto agar.

You need to make sure you have ready everything you need and that you work aseptically when carrying out the viable count. It is worth checking that the agar plates are fully dry before you start this work, because if they are wet it can affect the results and delay work. It is also a good idea to warm the plates to room temperature before you plate the cultures onto them.

To start, take the bacterial culture (working a blue Bunsen flame) and remove 1 mL of the culture. Carefully pipette this into 9 mL of sterile broth in a test tube or universal container (UC). Mix well (this is important; if you do not mix well, the results will be inaccurate). To mix, you can agitate the tube by hand or pipette or vortex the bacteria (Fig. 2.10). If using a vortex or mixing by hand, try not to get the culture in the lid of the culture vessel; otherwise, you might get it on your hand or drip it on the bench when you remove the lid.

Once it is mixed, take 1 mL of the culture and pipette it into 9 mL fresh sterile broth in the next sterile tube or UC. Continue doing this until you have dilutions that range from your concentrated sample to 10^{-7} (Fig. 2.11).

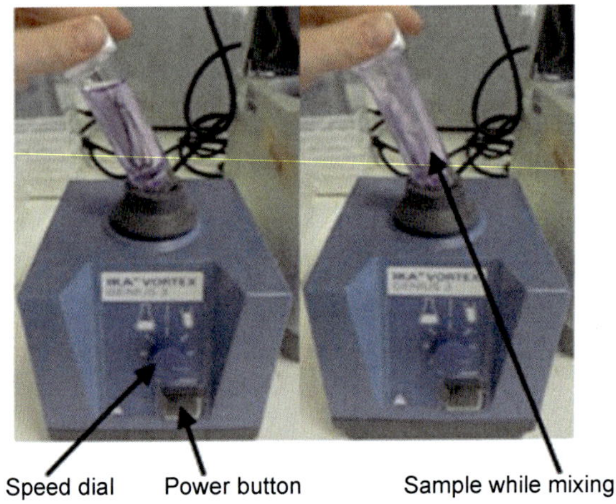

Figure 2.10 Using an IKA Vortex Genius 3 shaker to mix the sample before pipetting 1 mL of the sample into the next tube containing 9 mL sterile broth.

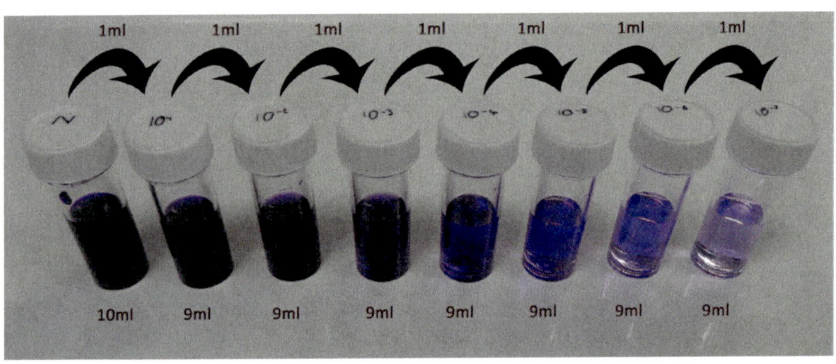

Figure 2.11 Serial dilution of bacteria (liquid has been coloured to highlight dilution effect). Bacteria are diluted from the neat overnight culture (far left) down to a 10^{-7} dilution (far right) by pipetting 1 mL of bacteria from each tube into the next. Each tube is mixed well before the bacteria are transferred.

You can then pipette each dilution onto a sterile agar plate divided into eight sections (Fig. 2.12). It is a good idea to prepare duplicate plates to ensure you get reproducible cell counts. It also means that if you contaminate a plate or have a wet plate that does not grow the bacteria properly, you might still get a usable count from the other plate. The volume you pipette will affect how long it takes your plate to dry, as well as the calculation you perform to determine the cfu per millilitres of the original sample (commonly volumes are 10–100 μL).

If you use a larger volume, you would need to use one agar plate per dilution and spread it on the plate (see subsequent description for the spread method).

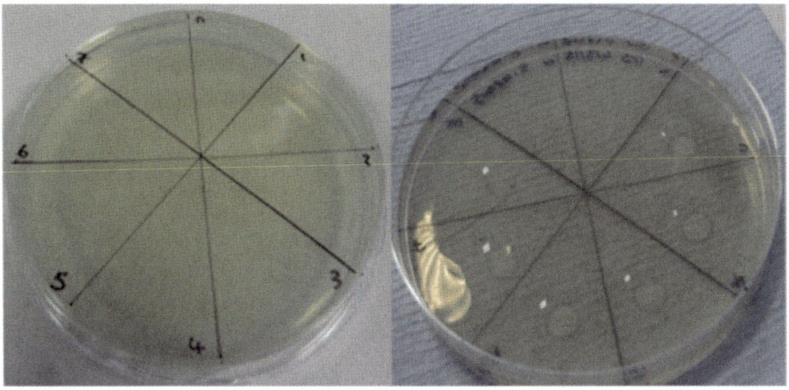

Plate with sections and dilutions marked on it

Diluted bacteria pipetted onto surface of agar, these need to dry before you invert the plate for incubation

Figure 2.12 Agar plates divided into eight sections of ready diluted bacteria to be pipetted into the corresponding sections. Labelling is on the bottom of the plate so that when bacteria are added to the surface of the agar, the numbers proceed (n to −7) in an anticlockwise orientation, opposite what you see when you label the plate.

Once you have pipetted the sample onto the sterile agar plate, wait for it to dry. Normally, this takes 5 or 10 min, depending on how dry the plates were before you started and how much you pipetted onto the plate. Once it is dry, invert the plate and incubate it overnight before removing it and counting the colonies.

To calculate cfu per millilitres:

$$\text{Cfu/mL} = \text{number of colonies} \times \text{dilution factor}$$

So, if you counted 45 colonies on a plate that had 50 μL of the 10^2 dilution pipetted onto it, you would need to calculate:

$$45 \times 10^2 \times 20 = \text{cfu/mL (you multiply by 20 to get from 50 μL to 1 mL)}$$

In reality, you will often find that you are working with much smaller volumes than in this example. It is common to scaled down the technique to fit the volume in a sterile Petri dish lid (Fig. 2.13).

If you want to try the smaller-scale version, it is still important to be sure each dilution is thoroughly mixed. If moving 20–180 μL (still a serial dilution), use the larger pipette (200 μL) to make sure each drop is well-mixed. If you use the 20-μL pipette, you will find that the drops do not mix properly. When using this method, there is also a risk of knocking against the Petri dish lid and causing all of the drops to run together. If this happens, you will need to start again. Remember to perform your work in triplicate to ensure that you are obtaining accurate results.

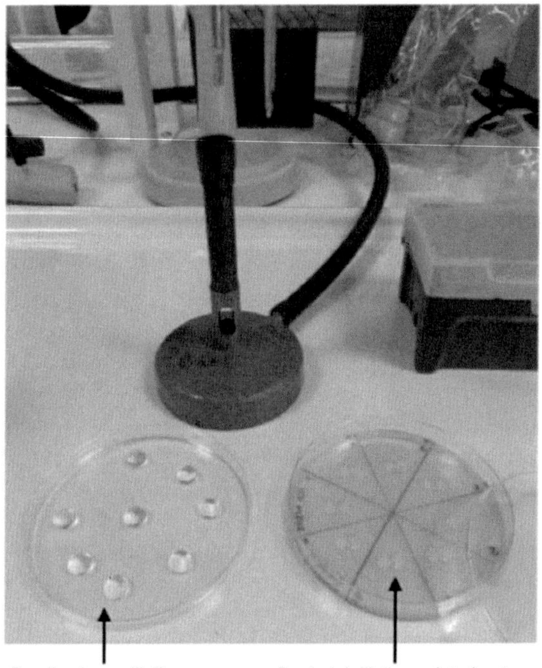

Figure 2.13 Petri dish lid containing small-volume dilutions of bacteria. These are then pipetted onto the surface in the same way as large-volume dilutions.

Common problems:

1. The most common issue was mentioned earlier; it results from not mixing the dilutions thoroughly enough before transferring liquid from it into the next tube. If this happens, you may see a higher cfu per millilitre in a lower dilution or large differences between replicates. If this is the case, you should repeat your work.
2. Another common issue occurs when plates are not dry enough before you start. You might find that when you pipette the bacteria onto the agar surface, they run to the edge of the plate or into another one. This normally results in colony counts that are not reproducible, and you should try again with drier plates.
3. You may find colonies growing in an odd place on the plate. This can be caused by accidentally releasing some of the culture from the pipette either before or after pipetting the culture on the plate. It can result from moving on from the drop too quickly. It will alter the count because the drops are unlikely to be in the right segment of the plate.

4. Counts may match up with plate labels. This is common when performing this procedure for the first time, and it happens because the bottom of the plate is labelled so that when it is flipped over to work, the dilution is the opposite way around. This is normally fairly obvious, and as long as you are sure that this is what you have done, you can work out which dilution is which and still take the counts.

Another way to get a viable cell count is to use a spread plate. To carry this out, you need to dilute the bacterial sample as described in Section 2.6. In this case, however, you use a whole plate for each dilution and can pipette a larger volume of the sample onto the plate. If you use this method, you will need more agar plates because you use one plate per dilution. Therefore, it is a good idea to plan ahead and make sure you have made enough agar to before you start, because you do not want to run out of plates partway through the work.

As with all the work you carry out, make sure you have everything ready before you start and label your plates with your name, date, the bacteria being plated and the dilution that will go onto that plate. You will need an L-shaped spreader to carry out this work. These can be plastic (and you should dispose of it after each use) or glass. The glass spreader would need to be sterilised after each use, normally by flaming with alcohol such as 95% ethanol. If you use ethanol, make sure you keep the volume small and have a fireproof cover for the container of alcohol in case you accidentally ignite it (see Chapter 1.3). You are most likely to ignite the alcohol by putting your spreader back into the alcohol when the spreader is too hot, so it is worth waiting for the spreader to cool before you sterilise it. To sterilise the spreader, dip the L-shaped part of the spreader and a small portion of the stem into the alcohol and then pass the spreader through the Bunsen flame. The alcohol will ignite and extinguish rapidly, but if you dip the spreader too far into the alcohol, you risk burning your hand, so be careful when doing this. The spreader will need to be sterilised between every sample or it could affect the results. It is important to let the spreader cool enough for you to use it to spread the bacteria without killing them. If you are unsure, touch the spreader to the sterile agar surface. If the agar makes a noise, the spreader is too hot.

When you have everything you need, take the undiluted sample of bacteria and make a range of dilutions from the original sample through to 10^{-7}. Continuing to work aseptically, pipette the samples onto each plate labelled with the corresponding dilution. Try to get the sample in the middle of the plate so that it is easier to spread it evenly over the plate. The volume of sample used is normally 100–200 μL. This can be spread evenly around the plate and will not take long to dry. It is important to start spreading quickly after you have pipetted the sample. You do not want it to start drying into the plate before you have had a chance to spread it. When making a spread plate, it is also important that the plates are completely dry; otherwise, the inoculum can take a long time to dry into the plate. Remember to use the same volume on every plate and record the volume you used, because you will need it to calculate cfu per millilitres after incubation. To spread the sample around the plate, take the L-shaped spreader and move it up and down the plate while turning the plate, to ensure you have spread the sample evenly. Some labs will have a rotating table that will turn the plate for you as you move the spreader up and down the plate. Be careful when spreading the sample; if you put too much pressure on the spreader, you are likely to damage the surface of the agar, which might affect the cell count. Once you have spread the sample onto the plate, let the plate dry (normally 5–10 min) before inverting it and incubating it at the appropriate temperature.

After incubation, to calculate the cfu per millilitres, pick the plate with the dilution that gives a count of 30–300 colonies. Each colony should be isolated and easily countable. More than 300 and they are likely to be touching on the plate, making an accurate count difficult. You can buy hand tally counters to keep track of the number of bacteria you have counted. Reset it between each plate count and record your observations in your laboratory notebook. Colony counters are also available on which you can place the agar plate. A light behind the plate makes the colonies more visible, and as you touch the lid of the agar plate with a pen it automatically registers the count. This can be useful because the pen marks on the lid can help you see which colonies you have already counted, so it is worth finding out whether a colony counter is available in your lab (Fig. 2.14).

You can also ascertain the total number of viable bacteria by using the pour plate method. For this technique, you will need molten agar (enough to fill a Petri dish approximately 18–20 mL). Make sure it is kept in a water bath at 50°C. If you keep it on your bench, it will become lumpy or completely solidify in its container while you are preparing your bench. Label sterile Petri dishes and prepare the bacterial dilutions as discussed earlier. Working aseptically, pipette the samples into the bottom of the empty sterile Petri dishes you have prepared. You can use any volume between 100 and 1000 μL (1000 μL is commonly used). It is best to keep the volume the same throughout your work and to record it in your lab book because you will need it to work out the final cfu per millilitres later.

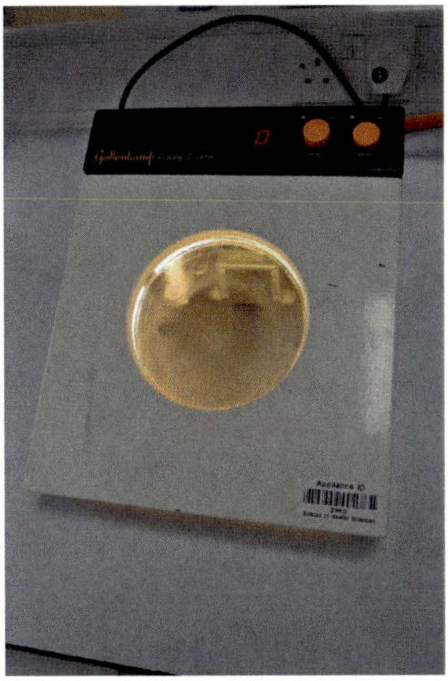

Figure 2.14 Colony counter with the back light on. If you place an agar plate over the light, it will aid you in counting the colonies. Note the counting window at the top. Each time your touch the plate or colony, the plate counter will record it, which enables you to make an accurate count.

Take the molten agar and transfer it aseptically to the Petri dish containing your sample. If you are pouring the agar from a glass container such as a UC, flame the top of the glass before pouring. Also, give the agar time to cool to just below 50°C before pouring (5 or 10min is normally sufficient). If you pour when the agar is too hot, you will kill some of the bacteria and alter the results. Once you have poured the agar into the Petri dish containing the sample, replace the lid and swirl the Petri dish gently to mix the sample. Be careful swirling the plate. If you are too rough, you will splash agar up the side of the dish or even onto the lid, which can stick the lid on and affect the count. The plate must then be left on the bench to solidify before you can incubate it. You should leave enough time for the plate to solidify before you move it. I it is half set when you move it, the surface of the agar may become lumpy, which can make it harder to count the colonies later. If you are making multiple plates, you will need to move the first plates away from the Bunsen as soon as you have poured and swirled them, to avoid this problem and make enough room for the later plates. Once the agar has solidified, invert the plate and incubate it at the appropriate temperature overnight.

Remove the plates from the incubator the following day and inspect them. They will have colonies growing both on and in the agar; you will need to count both types. Pick a dilution that gives 30–300 colonies and record the numbers in your laboratory notebook. The calculation is the same as that used previously; however, if you used a millilitre of sample in the bottom of your plate, you will only need to multiple the cell count by the dilution factor.

Direct microscopic counts of the bacteria can also be used to enumerate the bacteria in the sample. This can be achieved by using a counting chamber. These are specially designed microscope slides with a chamber that has an etched grid within it. This allows you to count the bacteria accurately in the sample using only a small volume of the sample. To use this method, you will need a counting chamber designed for bacteria (a commonly used one is the Petroff-Hausser counting chamber) and also a coverslip to place over the top of the counting chamber. Counting chambers and coverslips can be purchased from various suppliers including Sigma; Prosource Scientific and Thermo Fisher Scientific (https://www.sigmaaldrich.com/united-kingdom.html, http://www.psscientific.com/default.aspx and https://www.fishersci.co.uk/gb/en/home.html). The coverslips are not the same as those used routinely in the laboratory, because they are designed for use specifically with the counting chamber slide. They need to be heavy enough to remain in place once the bacterial suspension is added. Check to see whether your laboratory has this equipment and the guidelines that came with it before buying it.

When you are ready to count the bacteria, carefully place a coverslip over the top of the central well area in the slide. The coverslip will rest on the borders of the edge of the well in the centre of the slide. Once the coverslip is securely in place, use a sterile Pasteur pipette to transfer approximately 10 μL of the sample into one of the wells on the slide. You could also use a low-volume pipette to do this. Although you add the liquid containing the bacteria to the well in the slide, it will be drawn under the coverslip onto the grid (which you will view under a microscope) by capillary action. Do not add too much, because you do not want bacterial suspension leaking out onto the bench top. After you have added the liquid to the slide, it is best to wait a minute or two for the liquid to finish moving before you try to view the bacteria.

Place the slide on a microscope under a low-power objective. An ×40 objective is often used to locate the grid on the slide. You should be able to see the grid clearly when you look down the microscope lens. To count the bacteria, you may need to increase the magnification,

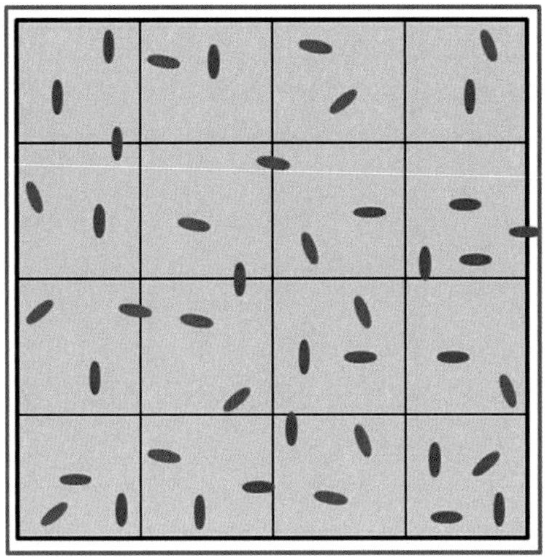

Figure 2.15 Simplified version of what you might see when observing the counting chamber under the microscope. Bacteria are represented in *purple*; the *red circle* highlights a bacterium, which is in two squares of the counting chamber. This demonstrates why you need a rule for counting bacteria that overlap boundaries.

because although the ×40 objective is powerful enough to locate the grid, you will not be able to see the bacteria clearly. Change the setting of the microscope to a ×100 oil immersion lens and check that you are focused on the grid and bacteria again (more details on how to set up microscopes are in Chapter 3). When looking at the bacteria, you should be able to see that they are evenly spread around the grid. If you have clumps, you will not be able to count the bacteria accurately. When counting the bacteria in each square of the grid, you also need a rule for bacteria that overlap the edges of the grid; otherwise, you may over- or undercount the bacteria (Fig. 2.15). It is probably best to draw a table in your lab book before starting and record the number of bacteria in each square as you go along. You can use the handheld tally counter to help you keep count.

To carry out the procedure accurately, you might need to make a dilution series of the bacteria, especially if the culture looks particularly turbid. If they overlap and are difficult to count, it is best to dilute the culture and start again.

Once you have a final count, you will need to use an equation to convert the counts to cfu per millilitres of the original culture. The equation you use depends on the volume of the counting chamber and whether you have diluted the sample. For example, if you used a counting chamber with a 0.02-mm-deep chamber and you counted all bacterial cells in a 1-mm^2 grid, you would use the equation:

$$\text{Cfu/mL of original culture} = \text{cells counted} \times 50,000 \left(\begin{array}{c} \text{this brings the} \\ \text{volume to 1 mL} \end{array} \right)$$

$$\text{e.g., cfu/mL} = 104 \times 50,000 = 5200000\,\text{cfu/mL}$$
$$5.2 \times 10^6\,\text{cfu/mL}$$

If you had diluted the bacteria before placing them in the chamber for counting, you would need to multiply that figure by the dilution factor, so if you had used a 10^2 dilution, the calculation would look like:

$$\begin{aligned} \text{Cfu/mL} &= (104 \times 50,000) \times 10^2 \\ &= 520000000\,\text{cfu/mL} \\ &\quad 5.2 \times 10^8\,\text{cfu/mL} \end{aligned}$$

For this calculation to work, you must know the volume of the area you have counted. Check this when you buy a cell counter or check with your laboratory manager if you already have a counting chamber that is used in the laboratory. It is best if you can check the original paperwork that came with the counter. Most counting chambers have information about the volume and grid size written on the slide, in which case you can easily check the volume. This method of counting your bacteria is similar to the spectrophotometer in the sense that this method will let you count both live and dead cells in a population rather than only the live cells that are in the pour plate, spread plate, and Miles and Misra count methods.

Now you should be equipped with the information needed to grow bacteria successfully in the laboratory, either in broth or on solid media. You have a starting point to help you if you need to identify an unknown sample. You should also be able to calculate the number of bacteria in a sample, which will help when moving forward with your experiments.

2.6 Notes page

Record observations and notes here.

CHAPTER THREE

Microscopy and staining

Contents

3.1 Introduction

The ability to view bacteria and stain-specific components of those bacteria is incredibly useful when working in bacteriology. Through visualising your bacteria, you can determine cell type (rod, cocci, spirochete, etc.) and through staining gain information about the basic composition of the cell wall. By using more advanced types of microscopy, you can also investigate cellular viability and cell surface or internal structural changes as well as look at the structure of bacterial communities. In this chapter, we will look closely at how to use a bright-field (light) microscope effectively and examine the uses for other microscopes such as dark-field, phase contrast and electron.

3.2 Light microscopy

The benchtop bright microscope is an invaluable piece of equipment in the laboratory. It is comparatively inexpensive and easy to use, and bacteria can be visualised with it after minimal preparation. If you are working within a microbiology department, you are almost certainly going to have one of these somewhere in the laboratory, so ask the laboratory manager where you can find it if it is not obvious. When handling your microscope, it is important to treat it gently because you do not want to damage any components. Therefore, if you need to move it around on the bench to get it into a better or more comfortable position for you to use, pick it up and move it, and do not drag it across the bench.

Bacteriology Methods for the Study of Infectious Diseases
ISBN 978-0-12-815222-5
https://doi.org/10.1016/B978-0-12-815222-5.00003-1

The bright-field microscope is a compound microscope (uses two lenses) that has one lens in the eyepiece and the objective lens. The microscope normally has a range of objective lenses that can be rotated into place above the stage, commonly ×4, ×10,×40 and ×100 objective lenses. These work in conjunction with the eyepiece lens, often ×10, to give you a maximal magnification of ×1000. This use of lenses allows greater magnification than would be possible if only one lens were used.

When using the microscope, it is important to understand the difference between magnification and resolution. Magnification allows you to increase the apparent size of an object, but if it is not resolved it will appear blurry. Resolution is the smallest distance at which two discrete objects can be separated and still be distinguished as two separate individual objects.

Before you start using the microscope, be sure you are familiar with all of its parts so you can adjust it effectively once you have a specimen to look at. Fig. 3.1 shows the basic structure of a light microscope and the parts that can by adjusted to improve what you see down the eyepieces.

If you are using a shared microscope or one that has not been used in a while, it is important to run a few basic checks before you start. There is normally a power switch on the microscope itself, and before you turn the microscope on, it is worth checking that the light on the microscope is turned down. Sometimes the light is left on at full power when the microscope is turned off or unplugged. If you turn it back on with the light on at full power, sometimes that can cause the bulb to explode. If you turn it down before you turn on the power, you can then turn it up as needed after the power is on.

After the power and light are on, look down the eyepieces. If you already see objects or shapes (before you have a slide on the stage), check both the eyepieces and the lenses on the microscope. Dirt, particularly fingerprints and mascara, may have been left on the eyepiece, and if the microscope has not been properly cleaned before being put away, there may be oil or dirt on the objective lenses. Make sure you use lens tissue to clean these parts because it is specially designed for use with the microscope and a different kind of tissue or cloth could lead to permanent scratches that will hinder your view when using the microscope.

Once you have a smudge-free view down the eyepiece, you should check the condenser and make sure that it is not completely closed, because this might make it hard to see anything when you first examine a slide.

It is essential that you know where the course and fine adjustments (Fig. 3.1A and B) are on the microscope you are using, because these are the

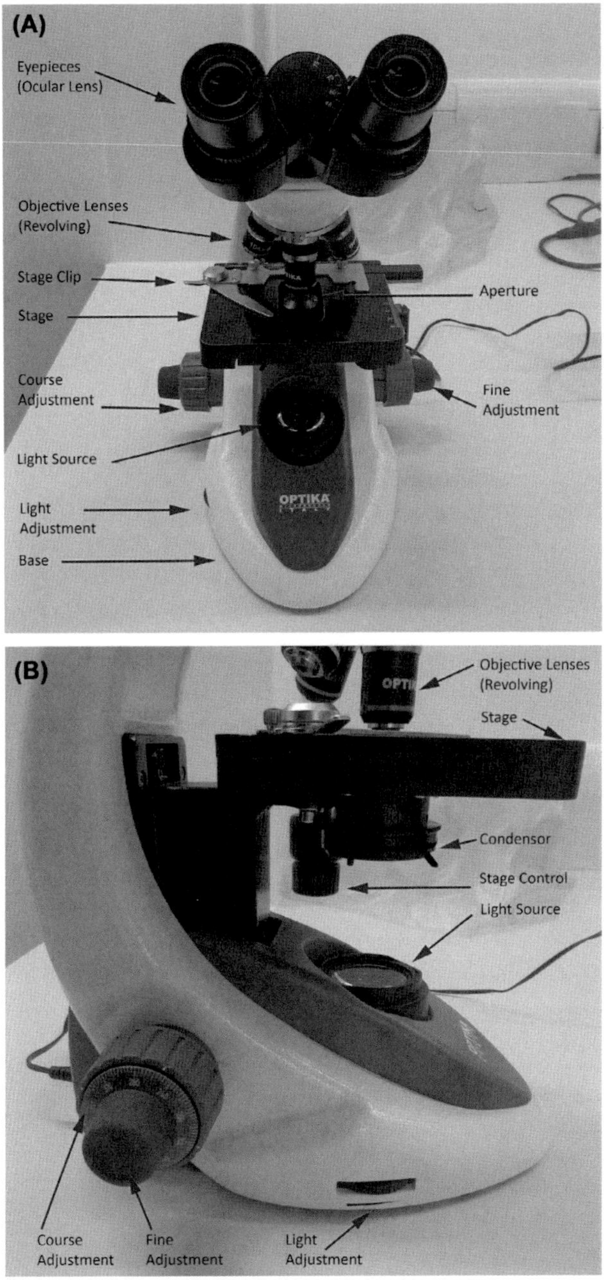

Figure 3.1 Example of a benchtop light microscope with important features labelled to allow you to use the microscope correctly and visualise your sample. (A) The microscope from the top and front. (B) The positions of those features below the stage or not easily visible from the first image.

controls that you will use most frequently. The course adjustment will move the body of the microscope up and down quickly and the fine adjustment will move the microscope up and down slowly. Before trying to look at the slide under the microscope, it is useful to move the fine adjustment to the centre of its range of movement. This gives you the scope to make small adjustments in both directions when observing bacteria.

3.3 Preparing slides

There are two basic ways in which you can prepare a bacterial sample for microscopy. The first is to use a wet mount or hanging drop method. Such a method does not involve fixing or staining cells, and so it can be useful if you want to examine live bacteria or motility or if you have a mixed sample that might contain other microbes such as algae or protozoa that you are interested in observing. The second way involves heat fixing bacteria to a slide. This allows you to stain the bacteria with various stains that can provide information about the bacteria; it also makes the bacteria easier to see under the microscope. Both types of method are outlined briefly next.

Hanging drop: To prepare a hanging drop slide, you will need a hanging drop slide (this is a slide with a concave depression in it), a coverslip, petroleum wax (cotton swab or small-volume plastic syringe), a small-volume pipette or inoculating loop and your sample. Make sure you have all these things clean and ready and prepare your microscope as described earlier.

Use the petroleum wax to create a ring around the depression in the hanging drop slide. This can be done by using a swab and placing the petroleum wax in a circle around the ring or by using a small-volume syringe to achieve the same effect. Once you have done this aseptically, transfer one drop of the sample to the coverslip, which should be on the bench in front of you. You do not need a lot of liquid when placing the drop on the slide; a loopful from an inoculating loop will give you enough to work with. Try to keep the drop as central as possible on the coverslip, because if it is too far to one side or near the edge, you will have trouble when you come to the next step. Once the drop is on the coverslip, take the hanging drop slide and place it onto the coverslip petroleum wax side down, with the depression in the slide above the drop. Take care not to touch the petroleum wax to the drop. If you squash the drop with the slide/petroleum wax, you will need to start again. If you have this problem, remember that both the slide and the coverslip have bacteria on them and need to be disposed of (coverslip) or cleaned (hanging drop slide) appropriately. When you have got the slide and coverslip together, push the slide down gently onto the coverslip so the wax forms a seal around the drop, then quickly and smoothly turn the slip over so the coverslip is on top. Take care when pushing the coverslip down so that you do not push too hard, because you do not want to smash the coverslip (they are delicate).

The sample drop should now be hanging down into the concave depression in the slide. It is important to make sure that you have made a seal with the paraffin wax so that if you are viewing it over a period of time, the liquid does not evaporate, but also that the sample does not leak out when you turn the slide over. Tips on how to visualise the slide under the microscope are covered in Section 3.5.

Wet mount: Another way to look at unstained live samples is to prepare a wet mount slide. This is a similar method that allows you to look at bacterial motility and does not require the use of the specialist hanging drop slide. To prepare this slide, you will need a clean microscope slide, a coverslip, petroleum wax (cotton swab or small-volume plastic syringe), a small-volume pipette or inoculating loop and your sample.

Set up your bench and microscope so that everything is ready and then start preparing your slide. Use the cotton swab or small-volume syringe filled with paraffin wax to create a wall of paraffin wax around the edge of the coverslip. The paraffin wax is going on the coverslip, not the microscope slide; this is so that the paraffin is in the right place when you put the coverslip and microscope slide together. You can place the paraffin wax on the microscope slide if you prefer, but make sure the paraffin wax is placed on the slide so that there is enough room for the drop of bacterial sample to fit into the middle and not so big that there is a gap around the edge when you put the coverslip and microscope slide together. Once you have put the paraffin wax on the coverslip, aseptically add a drop or loopful of the bacterial culture onto the coverslip in the middle of the area surrounded by the paraffin wax. Place the microscope slide on the top of the paraffin, making sure the paraffin wax seals around the slide, then quickly and smoothly turn the slide over so the coverslip is on top. The slide is now ready to be visualised under the microscope (see Section 3.5).

Heat fixed: This method fixes the bacteria you want to observe onto the glass microscope slide so that you can stain them. They are not live when you view them. To prepare your bacteria as a heat-fixed smear on a microscope slide, you are going to need microscope slides, a diamond point marker (for labelling slides) or frosted end slides and a pencil, inoculating loop and water or saline.

The heat-fixed bacterial sample can be prepared from liquid culture or from colonies grown on agar. If preparing from liquid culture, first label the slide using the diamond marker pen. This allows you to scratch a label onto the slide. It is probably best to use the initials of the bacteria (or a letter or number coding system) because there is not a lot of room on the slide. Make sure you know what the labels represent and make the marks at one end of the slide away from where the sample will be placed so they remain easy to read. Aseptically remove a loopful of culture from the bacterial sample and place it in the middle of the slide. Use the inoculating loop to spread the liquid out on the slide over an area of roughly 1×2 cm. You want the smear to be thinly spread across the slide so that when it initially air dries it is a pale white patch on the slide. If you are using a colony from an agar plate, place a loopful of sterile phosphate-buffered saline on the microscope slide. This gives something into which to spread the bacteria.

If you do not spread out the sample enough, the bacteria will be too concentrated, and when you come to visualise it you will not be able to distinguish individual bacteria or their characteristics. You now need to leave the slide to dry completely before you move on to the heat fixing stage. Do not be tempted to wave the slide around in the air to get it to dry more rapidly, because you do not want to aerosolise any of the bacteria accidentally. Once the slide is completely dry, you can heat fix the bacteria (this is an essential step; if you do not heat fix it, the bacteria will be washed off the slide when you stain or wash the slide). Take the slide and pass it through a blue Bunsen flame approximately three times. This should be done quickly. Make sure the part of the slide with the smear on it goes through the flame. If doing this by hand, take care not to burn yourself. If you move the slide through the flame too slowly, the glass will heat up rapidly. You can use microscope slide-holding forceps if they are available. Do not try to heat fix the slide in an orange-yellow flame because this will cause the slide to become sooty and will not fix the bacteria to the slide.

3.4 Stains

You can observe bacteria without staining them, but usually you will need some type of stain to gain a specific piece of information about the cell or to make it easier to see the cells. You are likely to come across a few different types of stains; they are grouped into simple and differential stains.

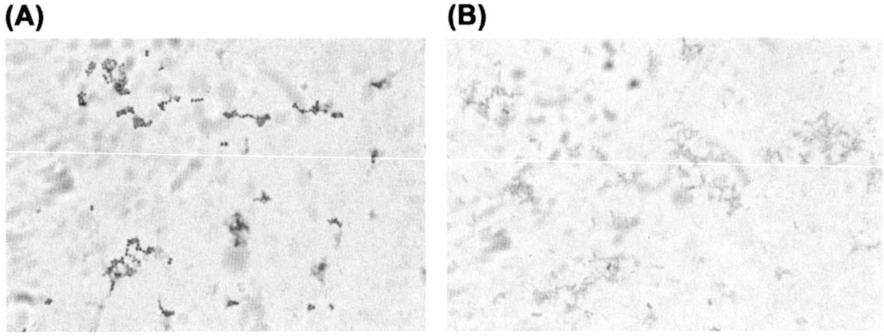

Figure 3.2 Magnified image of bacteria stained with methylene blue. (A) Cocci-shaped cells. (B) Rod-shaped cells.

Simple stain: This technique uses a single stain (e.g., methylene blue), which when applied to your bacterial cells makes it easier for you to distinguish cell shape, size and arrangement (Fig. 3.2) than if you were observing unstained cells.

Differential stain: This technique uses two contrasting stains that allow you to differentiate between one component of bacteria and another: for example, the different cell wall types in gram-positive or gram-negative bacteria.

One of the most useful and commonly used stains for which you will use during light microscopic observation is the Gram stain. The Gram stain allows you to differentiate quickly between gram-positive and gram-negative bacteria and makes it easy to decide on the cell shape (short rod, long rod, cocci, etc).

Before you undertake staining, there are some basics you will probably need, so before you start, make sure you have a bacterial sample, inoculating loops, microscope slides, a diamond pencil or frosted end slides and a pencil, gloves, a staining rack, water in a squeegee bottle, blotting paper, the stain and, in some cases, a detaining agent.

Gram stain: To undertake a Gram stain, heat fix the bacterial sample as described earlier and make sure you have Gram's crystal violet stain, Gram's safranin (or another counterstain), Gram's iodine, either 95% alcohol or acetone alcohol, water, blotting paper, gloves and a stop digital clock or a clock with a second hand. You will also need a staining rack and a tray or something similar in which to carry out the staining.

Do not try staining over a sink, because some of the stains used cannot be washed down the sink. Waste from the staining process will need to be collected and disposed of appropriately. You can set up the heat-fixed slides on a rack as shown in Fig. 3.3.

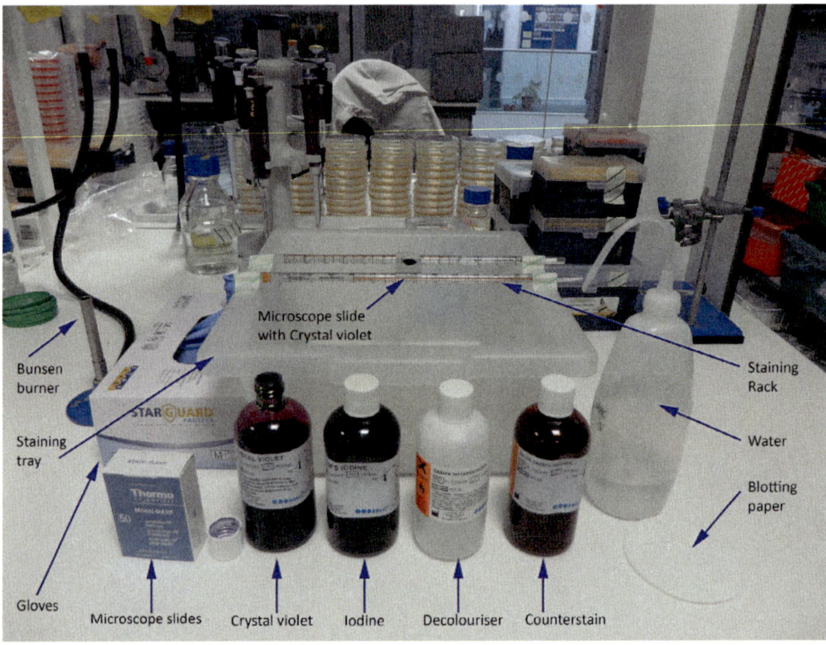

Figure 3.3 The equipment required to carry out a Gram stain procedure successfully using a staining tray on a laboratory benchtop.

When you are ready to start staining, make sure you have gloves on. If you get stain on your hands, it will not wash off. Take the crystal violet and flood the part of the slide where the bacteria are heat fixed. Make sure every part of the heat-fixed smear of bacteria is covered with stain. Leave the stain on for 1 min before tipping the excess off the slide and briefly rinsing with water. Do not spray the water directly onto the fixed bacteria because you could remove them from the slide. Add the water above the area of stain and let it run down over the stained bacteria. Next, add iodine over the same area as before and leave that for 1 min before rinsing with water in the same way as in the previous step. Then, tilt the slide and add the alcohol (decolouriser) onto the slide. You only need a small amount and it needs to be on the slide for only approximately 15 s. Once the liquid coming off the slide runs clear, rinse the slide with water. If you leave the decolouriser on the slide for too long or do not rinse it off quickly once the colour stops running, you risk removing the stain from all of the cells, not just gram-negative ones, so it is important step to get this step right. Once you have rinsed with water, lay the slide back onto the staining rack and add the counterstain (e.g., safranin or carbol fuchsin). Leave that on for 45 s before

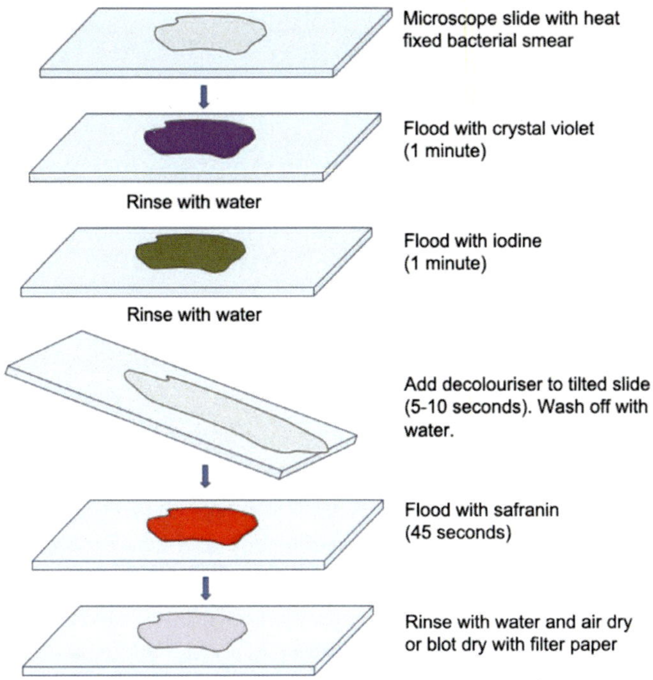

Figure 3.4 Diagram showing step-by-step guide to Gram stain procedure.

tipping the excess off and rinsing the slide with water (Fig. 3.4). You can either leave the slide to air dry at this point or move it onto blotting paper and gently pat the top of the slide with blotting paper to dry. Do not rub the slide with the blotting paper, or you will remove the bacteria. At this point, your slide is ready to be viewed (see Section 3.5).

Result: Gram-positive cells should look purple and gram-negative ones should look red-pink (Fig. 3.5).

If you are unfamiliar with the Gram stain, it can initially be difficult to decide whether you are looking at the gram-positive stain (purple) or gram-negative stain (red), especially if you only have one of these bacteria to look at and you are not sure what bacterium you have. To help with this, you can always stain a known gram-positive and a gram-negative organism at the same time as your test sample. This then means you should have a clear example of what each of the stain colours looks like under the microscope. You could use something like an *Staphylococcus epidermidis* or *Staphylococcus aureus* for a positive gram-positive cocci sample and *Escherichia coli* or *Pseudomonas aeruginosa* for a gram-negative rod sample. If you are still unsure, ask the laboratory manager to have a look through the microscope at your sample.

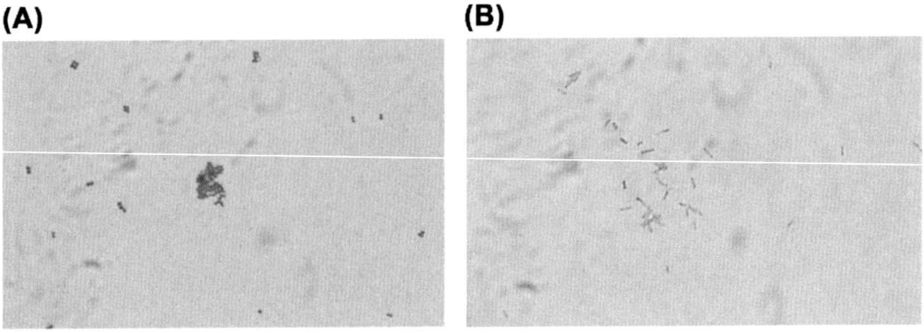

Figure 3.5 Images from a light microscope showing (A) gram-positive cocci and (B) gram-negative rods. Both were taken using a ×100 objective lens.

Methylene blue stain: Prepare the heat-fixed bacteria on the microscope slide as described in Section 3.3. Place the slide on a staining rack and cover the heat-fixed bacteria with methylene blue (make sure there is enough stain to cover all bacteria on the slide completely). Leave this stain on for up to 1.5 min before gently washing the stain off (do not aim the water directly onto the stain or you may wash the cells off). Once the water runs clear, place the slide on blotting paper and gently blot it dry. Do not rub the slide, or you will remove the cells. Once the slide is dry, it should be ready to visualise under the microscope (see Section 3.5).

Result: The bacteria on the slide should be stained blue.

Negative stain: This stain is used predominantly for bacteria that are hard to stain using other stains owing to the presence of a capsule. Commonly used stains for this technique include Indian ink and nigrosin. Congo red is also used sometimes. These can be purchased from suppliers such as Merck and Thermo Fisher Scientific (http://www.merckmillipore.com/GB/en). If you are working with bacteria that are often used in your laboratory group, check to see what stains they have available and whether they have made any changes to their methods (culture time, temperature and media can affect capsule formation). Using these stains will allow you to visualise the capsule and shape and arrangement of the bacteria that would be difficult to see without the stain. The stain will not attach to the bacteria, but instead stains the background, making the capsule or outline of the bacteria easy to observe.

To use the stain, with a Pasteur or low-volume pipette, place one drop of the stain at one end of the microscope slide. To this drop of stain add one loopful of bacterial culture and mix the two together. You want the cells to be well-mixed with no clumps visible but try to maintain the drop for the next step. After you have disposed of the inoculating loop, take a second microscope slide and place its short end onto the drop of stain/bacteria on the first slide. Move the second slide smoothly along the first slide, spreading the stain and bacteria along the slide as you go, creating a smear of cells on the first slide. Dispose of the second microscope slide and wait for the stain and bacteria on the first slide to air dry. Do not heat fix this slide. Heat can damage the bacterial capsule. This stain is air dried, and because the stain does not attach to the cells, it has the added benefit of not causing changes to bacterial morphology, which can occur when directly staining or heat-fixing bacteria. Once the slide has air dried, it is ready to visualise under the microscope. Because there was no heat fixation, the bacterial cells are live, so remember this when handling and disposing of the slide. If you are trying to observe capsule formation in the bacteria, you are more likely to see them in an old culture (more than 5 days old), because newly formed cells will not yet have a capsule.

If you want to stain cells that do not have a capsule or that have a mixed culture of cells that stain easily and those that do not, you could add a second stain after the slide has air dried. Once the negative stain has air dried, cover the area of the dried stain with a simple stain (such as methylene blue or Gram's crystal violet) and leave it for 1 min. To remove the crystal violet from the slide, tilt and allow the stain to run off. Do not wash the crystal violet off the slide or blot it because you will remove the bacteria that are not heat fixed to the slide.

Result: When you observe the slide through the microscope, you should see the background as a black colour. Any capsules around the cells as a transparent area and cells with no capsule will be stained purple, because they will have taken up the crystal violet stain.

Acid fast stain: There are several acid fast-staining techniques; one of the most frequently used methods is the Ziehl-Neelsen stain. You might see these terms used interchangeably or see staining reagents labelled with ZN rather than AF. Acid fast is a stain used for bacteria that have a particularly waxy cell wall. This waxy cell wall does not allow other stains discussed in this chapter to penetrate (bacteria for which you would need to use an acid fast stain include those from the *Mycobacterium* and Nocardia genera). With the acid fast procedure, the stain penetrates past lipids in the cell wall, where it is retained before destaining any other cells present in the sample. These are then counterstained with a different colour stain.

Prepare the microscope slide with bacteria as described in Section 3.3. Make sure with these bacterial samples that you thoroughly mix the cells with the inoculating loop on the microscope, because they tend to clump together, which can make imaging them more difficult. Once you have heat-fixed the sample, add carbol fuchsin to the slide. Place the slide onto a staining rack in a steaming water-bath and allow the slide to heat up (do not let the slide enter the water). The stain on the slide should steam for 5 min. If the stain dries out, make sure you add more carbol fuchsin.

If you do not have a water-bath (a beaker of water on a hot plate will work as well), you can perform the heating step by placing blotting paper over the heat-fixed stain and saturating it in carbol fuchsin. The slide can then be heated gently in a Bunsen flame, but you must be careful not to let the paper dry out and catch fire. The stain-soaked paper should steam, but if it looks as if it has stopped steaming or is drying out, add more carbol fuchsin to the paper.

After 5 min, remove the slide from the heat and allow it to cool for a couple of minutes. Rinse with water (do not put cold water straight onto the hot slide or it might crack) and add the destaining agent (acid-alcohol). If you tilt the slide when adding the destaining agent, you will be able to see when it starts to run clear. Rinse the slide again with water. At this point, you will need to add the counterstain. Methylene blue is commonly used. Add the counterstain to the slide and leave it to stain. After a couple of minutes, rinse the counterstain off with water and dry the slide with blotting paper. Blot, but do not rub the slide or you will lose the cells. The slide should now be ready to visualise under a microscope (see Section 3.5).

Result: When visualised, the slide should show the acid fast bacteria as red cells; any other cells in the sample will be stained blue.

Spore stain: The spore stain can be useful if you are working with bacteria that produce spores or you have an unknown sample and suspect that this might be the case. To undertake a spore stain, heat fix the bacterial sample as described earlier and make sure you have malachite green, safranin, water, blotting paper, glove, a clock and a waterbath or beaker filled with malachite green and a Bunsen.

Prepare the bacterial smear as in Section 3.3. Place a piece of blotting paper over the smear (make sure that the blotting paper does not hang over the edges of the slide or you risk setting it on fire). Soak the blotting paper in malachite green and then place the slide over a Bunsen flame. You can do this by placing the slide above a boiling water-bath or beaker on a hot plate. The slide can be held on a staining rack or with microscope forceps if you have only one slide but be careful not to burn yourself with the steam.

The slide can also be held above a Bunsen flame with either method. You should see steam rising from the malachite green/blotting paper; this needs to steam for around 5 min. If it stops steaming, add more malachite green to the slide/blotting paper but do not add so much that it runs off the slide or cools the slide to the point where it stops steaming.

When the stain has been steamed onto the slide for 5 min, remove the slide onto clean blotting paper. Remove the original blotting paper from the top of the slide and dispose of it and allow the slide to cool. Once the slide is at room temperature (this takes a couple of minutes), rinse the malachite green from the slide using water. When the water runs clear, add safranin to the slide and leave it for 2 min. Rinse the slide again with water and blot it dry. Do not rub the slide, or you will remove the cells. The slide is now ready to be viewed with a microscope (see Section 3.5).

Result: Once they are stained, you should see red bacterial cells (vegetative cells) and green spores (endospores). The spores can be located in several places in the cell (Fig. 3.6), either placed at the centre, at the end or between the centre and the end of the cell.

You will also normally see spores that are free from vegetative bacterial cells. The placement of the spore within the cell will often help if you are trying to identify the bacteria in the sample. Some medically important spore-forming bacteria are listed in Table 3.1.

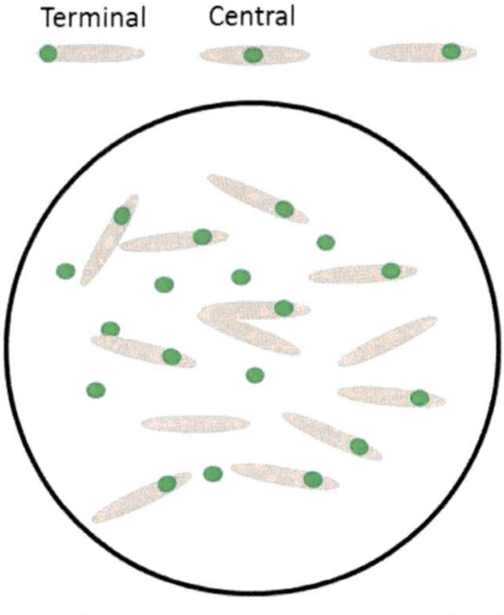

Figure 3.6 Representation of what spore-stained bacteria visualised using a benchtop light microscope will look like if the described protocol is followed.

Table 3.1 Examples of spore-forming bacteria of medical importance.

Bacterial name	**Spore location**
Clostridium difficile	Subterminal
Clostridium perfringens	Subterminal
Clostridium botulinum	Subterminal
Clostridium tetani	Terminal
Clostridium novyi	Central or subterminal
Bacillus cereus	Central
Bacillus anthracis	Central

3.5 Using the light microscope

At this point, you should have a microscope that is ready to be used and bacteria that are heat fixed and stained and ready to be visualised under the microscope. The first thing you will need to do is to place the slide on the microscope viewing platform and ensure that it is in the correct place so that it moves properly when you want to look at the slide. Most microscopes will have a stage clip or specimen holder to ensure the microscope slide stays in the right place while you are viewing it (Fig. 3.7).

Once the slide is in the correct place, you will need to select an objective lens with which to view the slide. It is usual to start with the lower-power objective lens (typically the ×10 objective). Make sure that you start with the objective lens over the centre of the microscope slide and over the centre of where you have stained (cells), to give yourself the best chance of finding the bacteria on the slide. It is also essential to make sure that the slide is the correct way up; if you accidently place the slide upside down, you will find it difficult to see the bacteria. Once you have selected the objective lens, start with the lens as close to the slide as possible and use the course focus to move slowly away from the slide. As you move away, you should see the bacteria come into focus. It is common to move the lens too far and go past the bacteria. If this occurs, you might notice them as a flicker of different colour across your vision. If you manage to bring the cells successfully into view, you will need to switch to the fine focus to get a sharply focused image of them. Using the lower objective lenses, you should be able to find the bacteria. This will allow you to move the slide so that the bacteria are in the middle of your field of vision. This is important because, as you move to a higher magnification, your field of view will decrease so that anything on the edges of your field of view at

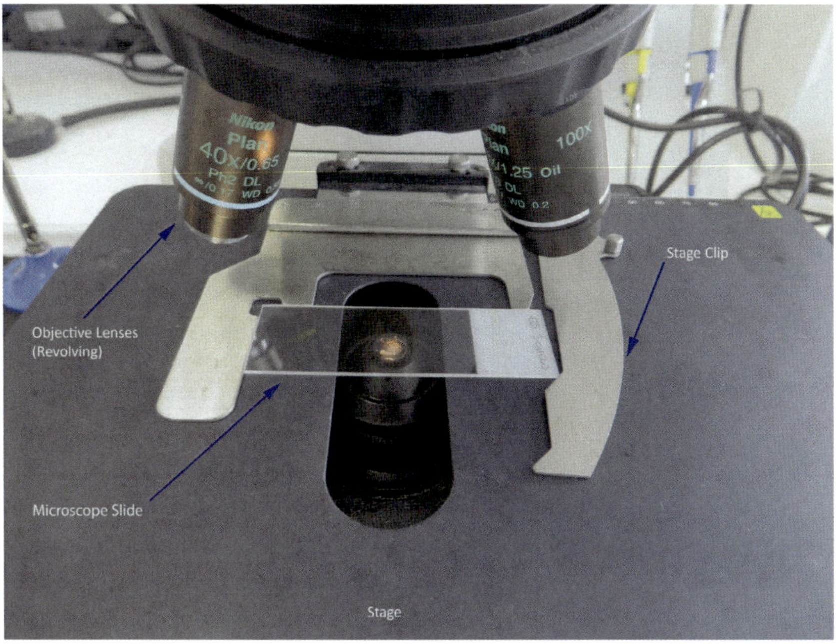

Figure 3.7 Microscope stage with a microscope slide held in the right place on the stage to enable bacterial cells to be viewed. Note that the objective lenses have been moved out of the way while placing the slide on the stage. To view the cells, a lens would need to be swung into position above the slide.

a ×10 objective lens will be missing when you go up to a ×40 objective lens. You will not be able to see any level of detail at the lowest power objective lens, so once you have the bacteria central and in focus with the lower power objective lens, move to a higher power objective lens. Often, the next objective lens up is a ×40 objective. You should be able to move the ×40 lens into place over the stage by adjusting the course or fine adjustment and be able to see the cells. Use the fine focus to adjust for a sharper image. If you cannot see the bacteria when you move the higher power objective lens around over the slide, you can use the course adjustment to move the ×40 objective to its lowest point (closest to the slide) and then work away from the slide slowly to bring the cells into focus. If you want to see details of the bacterial cells, you will need to use the ×100 oil immersion lens. To do this, move the ×40 objective away from the slide and place a small drop of oil onto the part of the slide that is central and will be under the lens when you are viewing the slide. Swivel the ×100 lens around to the slide and the tip of the lens should touch the oil/slide.

It is important when working with the ×100 oil immersion lens that you use only the course adjustment to move the lens away from the slide when you look through the eyepiece. If you move the course focus toward the slide when looking through the eyepiece, you could accidently smash either the coverslip or the microscope slide if you take the objective lens too far down. You will need to move the focus very little when using this lens; if the lens comes out of the oil, you will have moved the lens too far and need to start again. Do not forget when moving up to this magnification that if you do not have the cells you want to look at in the middle of the slide before you increase the magnification, you might not see them in the new smaller field of view after you have increased the magnification, so always centralise the object of interest on your slide before increasing the magnification. Once the cells are in focus, you can then change your field of view by using the stage control to move the slide around the stage.

You can also use the microscope in conjunction with a stage micrometer and an eyepiece graticule to work out the size of the bacteria in the sample. The stage micrometer is etched onto a specialist microscope slide and has a precise scale. Different sizes and scales are available; often they are 1 mm long and are divided into 100 sections that are 0.01 mm (10 μm) each. As you change the magnification, the apparent size of the stage graticule will change but the eyepiece graticule will stay the same. Thus, if you know how many micrometres each of your eyepiece divisions represents at each magnification, you can use it to measure your cells. You can get different-length eyepiece graticules, so check before you start working with yours; commonly, they are 10 mm long, so that will be used in this example.

First, remove the eyepiece lens and place the eyepiece graticule (a small round glass disc with the scale on it) in the lens space. Before replacing the lens on top, look down the eyepiece and make sure the micrometer is the right way up. Then, place the stage graticule onto the microscope stage and view it at the magnification that you are going to use to view your cells. You should calibrate for every magnification you will use.

Make sure you have the stage micrometer and eyepiece graticule lined up before you try to count the divisions (Fig. 3.8). Once they are lined up, you can start your calculations. For example, if, as in Fig. 3.9, you have 52 eyepiece units to 10 stage graticule units, it means that 45 eyepiece units are 100 μm long (only at that magnification); therefore, one eyepiece unit is (100/52) = 1.9 μm. You can now replace the stage micrometer with the sample at this magnification and measure the size of the cells. If you change

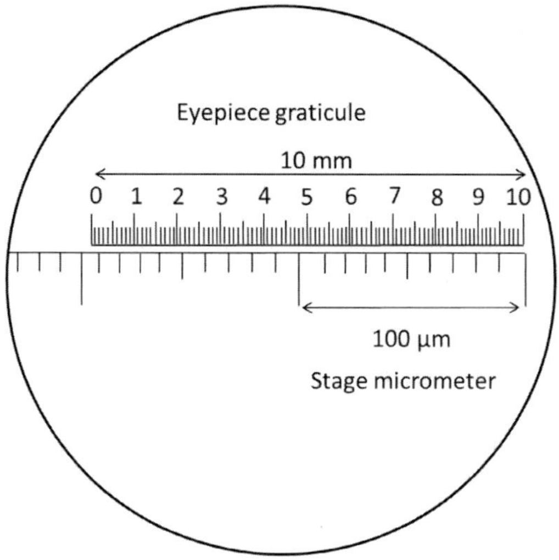

Figure 3.8 Diagrammatical representation of what will be observed with the microscope when using both the stage micrometer and eyepiece graticule to calibrate the measurements at a particular magnification.

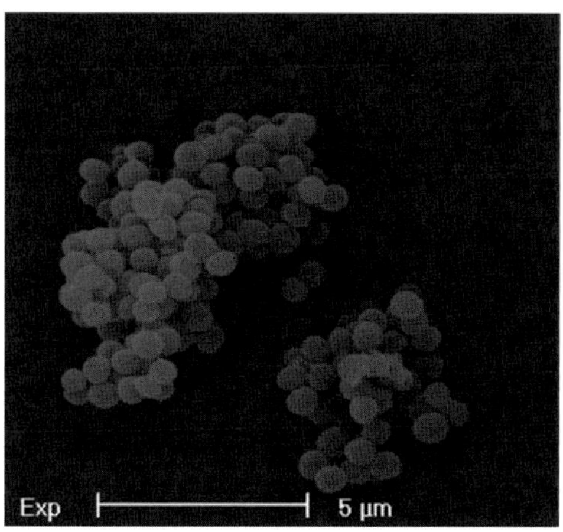

Figure 3.9 Scanning microscopy allows you to see the surface of bacteria. It also makes it possible to see the structures that bacteria make when grouped together. Here, clumps of *Staphylococcus aureus* are observed after growth in nutrient broth.

magnification, you will need to recalibrate to make sure you get an accurate measurement. The measurements can change between microscopes even when you use them at the same magnification, so if you change microscopes at any point, make sure you recalibrate. Do not forget that as you increase your magnification you might need to move the stage micrometer around to ensure it is in the centre of your field of view and in line with your eyepiece graticule as you increase the magnification.

3.6 Other types of microscopy

Other types of microscope are available that allow you to look at bacterial cells in different ways. Some allow you to see the bacterial cell in more detail using higher magnification; some in conjunction with specialised stains allow you to see whether cells are dead or alive, pinpoint specific bacterial structures or look inside cells. These other types of microscopy require more complex preparation of bacterial cells and often use more expensive and complicated microscopes. The uses of these other types of microscopes will be briefly outlined subsequently and should give you an idea of which one you might want to look at in more detail if you need something in addition to the benchtop light microscope.

Electron microscopy is a powerful tool for obtaining highly magnified images of bacteria. Instead of using light to image the bacteria, this form of microscopy uses a beam of electrons and allows you to see fine details of the bacterial cell without losing resolution. There are two main types of electron microscopy: scanning electron and transmission electron (there are other types as well). Scanning electron microscopy allows you to see the outside structure of bacteria in detail (Fig. 3.9) and can be used, for example, to see whether an antimicrobial agent has an impact on cell structure or integrity. It also allows you to see the structure of bacterial communities, such as how more than one type of bacterium group together if you have grown mixed bacterial culture or whether your bacteria have stuck to a certain type of material or surface.

Transmission electron microscopy allows you to look at a section sliced through the bacteria to observe the structures inside bacterial cells. It can be used to look at structures such as the nucleoid, flagella, or septa (Fig. 3.10).

If you are going to undertake electron microscopy, you will need electron microscopes. These are much larger than a benchtop light microscope and expensive to buy and maintain. If you have one of these microscopes where you work, you will need to be trained on how to use it, as well as how to prepare bacteria for imaging. The preparation of cells for these types of microscopy is far more complicated than the staining methods described previously and may involve, fixing, dehydrating, embedding, sputter coating, sectioning and negative staining, depending on which microscope you use. If this type of imaging happens regularly in your laboratory, ask about the

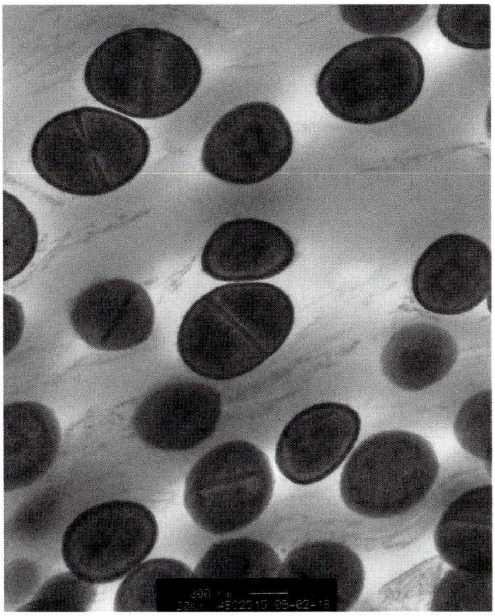

Figure 3.10 *Staphylococcus aureus* cells seen here show clearly defined cell walls with fully and partially formed septa visible across some of the cells.

protocol for preparing the bacterial cells. Proper preparation of the cells is essential for obtaining a clear image. If you have to use an electron microscope on another site (this is common), ask whether the researchers have standard protocols with which you can start. You can always optimise your protocols once you are familiar with how the process works.

With the information provided here, you should be ready to prepare bacterial samples for visualisation appropriately and to use basic microscopy to gather information about your samples. You should be able to adjust the microscope settings to get the best view of your samples, and the skills gained using a basic microscope should be useful if you decide to move on to more complex microscopy techniques.

3.7 Notes page

Record observations and notes here.

CHAPTER FOUR

Antimicrobial testing

Contents

4.1 Introduction

Antimicrobial agents have been used extensively across the world to inhibit microbial growth in a range of situations including medical, agricultural and environmental purposes. An understanding of how antibiotics target different sites within bacteria, such as cell wall components and protein or RNA synthesis among other targets has allowed different classes of antibiotic to be used for varying functions. Some are used for their broad range of activity, whereas others are employed for their narrow range of activity and ability to inhibit or kill particular types of bacteria.

The heavy use of antibiotics over many years has led to a rise in antibiotic-resistant bacteria and a corresponding decrease in the efficacy of many antibiotics. Therefore, the ability to determine the sensitivity of bacteria to antibacterial compounds is an essential technique that can be undertaken in a variety of ways.

4.1.1 Antibiotics and targets

The activity antibiotics have falls into two broad categories: bactericidal (they kill the bacteria) or bacteriostatic (they inhibit bacterial growth). The effect they have depends on the part of the bacteria they target. A wide range of antibiotic classes have been developed targeting varying parts of the bacterial cell. Examples of some antibiotic classes and their bacterial targets are given in Table 4.1.

Ideally, an antibiotic will display selective toxicity and target a component that is unique to the bacteria and will not cause damage to human

Bacteriology Methods for the Study of Infectious Diseases
ISBN 978-0-12-815222-5
https://doi.org/10.1016/B978-0-12-815222-5.00004-3

Table 4.1 Examples of some antibiotic classes with their corresponding target sites and known resistance mechanisms.

Antibiotic class	Example antibiotic	Target site	Common resistance mechanisms
B-Lactams	Penicillin	Cell wall	• Enzymatic degradation • Altered penicillin binding protein
Tetracyclines	Tetracycline	Protein synthesis (30s)	• Efflux pumps • Altered target site
Aminoglycosides	Gentamicin	Protein synthesis (30s)	• Enzymatic modification • Altered target site • Decreased uptake
Lincosamides	Clindamycin	Protein synthesis (50s)	• Altered target site
Glycopeptides	Vancomycin	Cell wall	• Altered target site
Polymyxins	Colistin	Cytoplasmic membrane	• Altered target site
Macrolides	Erythromycin	Protein synthesis (50s)	• Efflux pumps • Altered target site
Quinolones	Ciprofloxacin	DNA gyrase	• Efflux pumps • Altered target site
Rifampicins	Rifampicin	RNA synthesis	• Altered target site

cells. In practice, many antibiotics will have some level of host toxicity in addition to their antibacterial effect. There are several ways in which you can test the efficacy of antibiotics against the bacterial isolates in which you are interested. Once familiar with standard testing methods, those methods can also be adapted to test new compounds or materials that you may have developed to inhibit or kill bacteria. Commonly, these methods include testing the activity of antibiotics against bacteria on agar plates or in broth. A range of methods that can be used are outlined in this chapter.

4.1.2 Disc diffusion

One of the easiest and quickest methods that can be used to test the antibiotic sensitivity of a bacterial isolate is disc diffusion. When testing a conventional antibiotic, this involves creating a lawn of bacteria on an agar plate and then placing an antibiotic disc or discs onto the lawn before incubation it overnight. This method works because the antibiotic in the disc diffuses out into the agar, creating a concentration gradient. The concentration of antibiotic near the disc is high and gradually decreases the farther from the disc it diffuses.

The zone of inhibition caused by the antibiotic is measured the next day, giving an idea of how effective the antibiotic is against a particular bacterium. Standardised methods are available that provide detailed instructions on how to carry this process out in reproducibly: for example, from the European Committee on Antimicrobial Susceptibility Testing (EUCAST) (http://www.eucast.org/) or the Clinical and Laboratory Standards Institute (CLSI). These organisations provide detailed methods describing how to perform standardised testing, including information about which concentration antibiotic discs to use and what the breakpoints are for a variety of bacteria. When using the disc diffusion method, the breakpoint is the size of the zone of inhibition below which a bacterium is resistant. The organisations also provide the size of zone above which the bacteria are considered to be sensitive. If you are working with antibiotics, it is worth following these guidelines because they ensure that you are creating reproducible tests that can be compared with antibiotic sensitivity tests carried out in other laboratories. They provide details about how to prepare the media and inoculum, as well as information about how to calibrate and validate your work. These guidelines and breakpoints are updated regularly, so you should always check for the latest updates before performing this type of experiment.

If you are using the disc diffusion method with the intention of checking results against the breakpoint tables, it is important to check that the antibiotic concentration in the disc is correct, because some antibiotics have more than one concentration available. It is also useful to familiarise yourself with the antibiotic abbreviations, because not all of them are immediately obvious as short versions of the original antibiotic name (Fig. 4.1).

A general overview of the method is given here, but you should check the protocols provided by the organisations discussed, to ensure that the specific details are correct for the bacteria and antibiotic being tested.

The antibiotic disc susceptibility testing method is briefly:

1. Aseptically pick a few bacterial colonies from an overnight agar plate culture and suspend these in sterile saline to give a density equivalent to a 0.5 McFarland standard. You can do this using a spectrophotometer or by comparing the density with a standard if you have one in the laboratory.
2. When the suspension is at the correct turbidity, take a sterile cotton swab and dip it in the suspension. Press it against the inside of the suspension tube to remove excess liquid. Then, apply it to an agar plate. The swab should be swabbed across the plate, moving from side to side

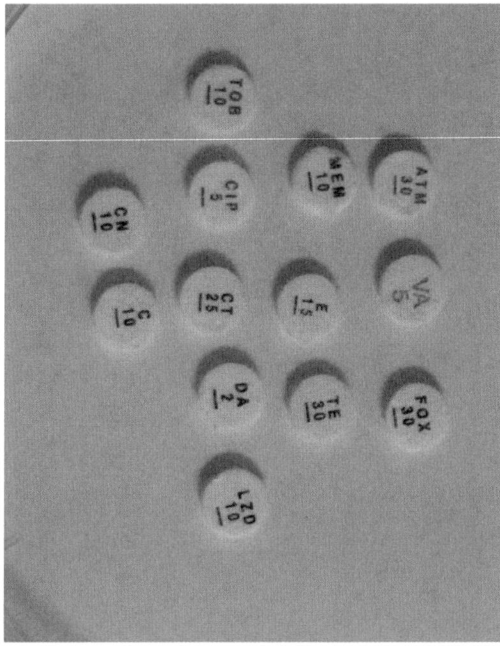

Figure 4.1 Range of antibiotic discs available for sensitivity testing. Letters on the discs give the identity and concentration in micrograms of the antibiotic in the disc. *ATM*, aztreonam; *C*, chloramphenicol; *CIP*, ciprofloxacin; *CN*, gentamicin; *CT*, colistin; *DA*, clindamycin; *E*, erythromycin; *FOX*, cefoxitin; *LZD*, linezolid; *MEM*, meropenem; *TE*, tetracycline; *TOB*, tobramycin; *VA*, vancomycin.

all the way from the top to the bottom of the plate before rotating the plate approximately 60 degrees. Before swabbing again, rotate the plate once more and swab again. The plate should then have been swabbed in three directions across the plate. The swab should be dipped into the inoculum only before the swabbing starts, not between each rotation.

3. Once the lawn of bacteria has been created, the antibiotic discs can be applied to the plate. These should be removed from the fridge or freezer before use and allowed to come to room temperature before use. The discs can be applied using an antibiotic disc dispensing machine if you have one in your laboratory; use sterile forceps if you do not have a dispenser (Fig. 4.2). If using forceps, take care not to rip or damage the disc when handling it.
4. Ensure that you press the disc down onto the agar so that it is secure and does not fall off. Then, invert the plate for incubation. Incubate the plates for 16–20 h at 35°C.

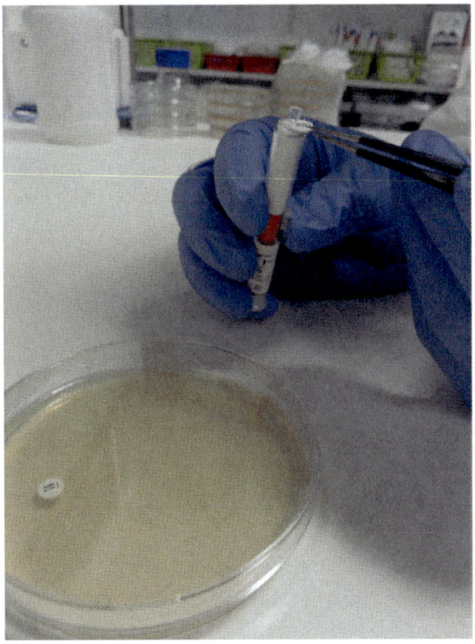

Figure 4.2 Forceps used to remove an antibiotic disc from its tube. The discs will come out of the tube only in one direction. If a disc is damaged because it has been forced out of the tube in the wrong direction, dispose of it and remove one in the correct direction.

If using forceps, these can sterilised between uses by dipping the tip of the forceps into ethanol and passing through a blue flame. Hold the forceps away from the flame and allow the flame to go out. Wait a few seconds for the forceps to cool before using it. Do not dip the forceps into the ethanol when it is hot, or the ethanol could catch fire. If in doubt, wait slightly longer before dipping the forceps. It is a good idea to have something nonflammable, such as a metal lid, that can be put across the top of the container, which holds the ethanol. If the ethanol catches fire, the lid can be placed across the top to remove the oxygen supply to the fire.

When undertaking disc diffusion testing, it is important to get the antibiotic discs onto the plates within 15 min of creating the bacterial lawn. After applying the antibiotic discs, you should aim to get the plates into the incubator within 15 min. So, think carefully when setting up your experiments, because you might need to prepare the lawn plates and apply the antibiotic discs in batches so that you do not stray outside those time limits.

To determine whether the bacteria are sensitive to the antibiotic discs you have placed on the agar, you will need to measure the zones of inhibition around the discs from the back of the plate. You can measure the zones using digital or manual callipers or a ruler. The measurement should be

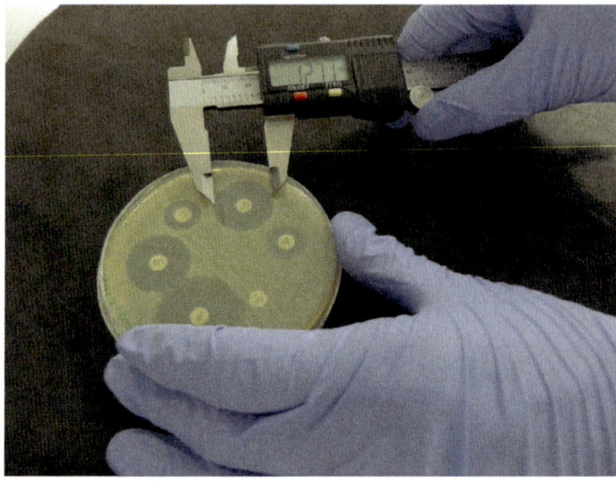

Figure 4.3 Digital callipers used to measure the zone of inhibition in millimetres around each antibiotic disc on the agar plate.

Table 4.2 Sample layout for a table to collect zone inhibition measurements.

	Antibiotic discs (μg/mL)					
Bacteria	**CIP (5)**	**CN (10)**	**TOB (10)**	**ERY (15)**	**C (2)**	**TET (30)**
MRSA1						
MRSA2						
MRSA3						
MRSA1						
MRSA2						
MRSA3						
MRSA1						
MRSA2						
MRSA3						

This table allows triplicate data for each organism to be collected. Some of the antibiotic disc abbreviations are easy to remember (e.g., TET for tetracycline) but others are less straightforward (e.g., CN for gentamicin). *C*, chloramphenicol; *CIP*, ciprofloxacin; *ERY, Erythromycin; MRSA*, Methicillin-resistant *Staphylococcus aureus*; *TOB*, tobramycin.

the diameter of the zone of inhibition (where no bacteria have grown) to the nearest millimetre (Fig. 4.3). The zones should be read against a dark background using reflected light, or from the front using reflected light. Whichever is used depends on the agar employed, so check the published guidelines to determine which is needed.

It helps to have a pre-prepared table in your lab book in which to record your results, especially if you have multiple isolates and antibiotics. A pre-prepared table ensures that the data are recorded accurately. Table 4.2 shows a sample data collection table that could be used to record zone of inhibition measurements from triplicate isolates.

Once the zone of inhibition is established, the sensitivity or resistance status of the bacteria can be determined by comparing that value with published breakpoint tables. It is good practice to run a control bacterium alongside the test bacteria. Testing a bacterial isolate with a known antibiotic susceptibility profile alongside the test bacteria allows you to check that the method and antibiotics are working as they should be and that the results obtained are reliable.

When inspecting the zones of inhibition, if they do not appear to be straightforward (e.g., double zones, fuzzy-edged zones or colonies growing within the inhibition zone), often detailed standardised protocols will offer advice on how these should be interpreted. It is worth checking the detailed protocols if there are unusual results, because advice on how to interpret the varying zones can differ among species.

Analysis: Once the data have been collected, they can be processed and presented as a bar graph to make it easy for others to see the results quickly. If the tests have been carried out in triplicate, there will be multiple values for the zone of inhibition for each antibiotic. If tested in triplicate, the mean value should be calculated. This can be done by adding together the zone of inhibition measurements for each antibiotic and dividing by the number of replicates; this will generate one mean value for each antibiotic, which can be plotted onto a graph. The standard deviation should also be calculated for each antibiotic to ensure that replicate data are similar. The standard deviation can then be added to the graph as error bars (Fig. 4.4). The standard deviation can easily be calculated and often is available as a function in a programmes such as Excel. To calculate standard deviation in a programme such as Excel, select an empty cell, press the Equals key, and then select standard deviation from the Functions menu bar (often the top left in Excel and displayed as STDEV or a variation of this). Select all replicates for the antibiotic of interest and then press Return. This should generate a value for the standard deviation. This process can be repeated for each antibiotic, and once generated, can be plotted onto the graph. The error bars should provide an easy way to see whether the zone sizes were similar between replicates or whether there is a lot of variation, which could indicate that the experiment needs to be repeated.

Depending on what is being tested, there are two ways in which the data could be presented. If you are testing multiple isolates of the same species and the same range of antibiotics on each isolate, a graph that contains all the isolates and only one antibiotic could be created. This type of graph provides an opportunity for the breakpoint line plotted onto the graph so

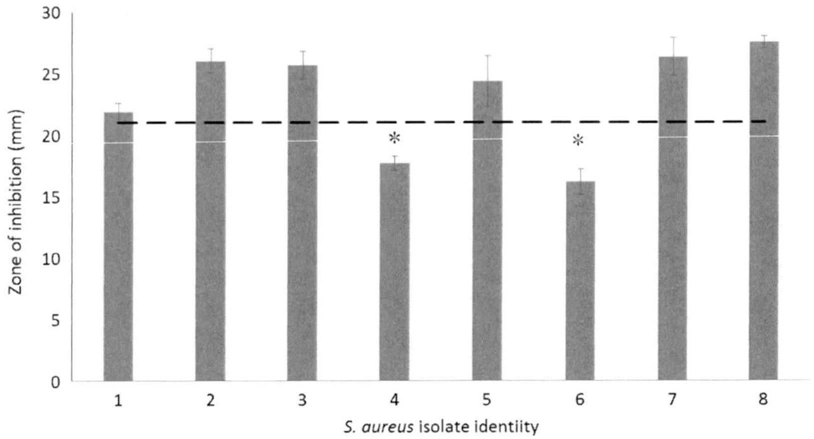

Figure 4.4 Zones of inhibition (in millimetres) generated by for ciprofloxacin against a range of *Staphylococcus aureus* isolates. The *dashed black horizontal line* represents the breakpoint beneath which isolates are resistant (marked with *asterisks*). Error bars represent standard deviations of the mean.

that anyone looking at the graph has an easy way to see which isolates are sensitive and which are resistant (Fig. 4.4).

If you are working with different species of bacteria and a range of antibiotics, it might be better to plot one bacterial species per graph with the whole range of antibiotics shown. Breakpoint lines can still be plotted onto this type of graph, but they will vary depending on which antibiotic and bacteria are being presented (Fig. 4.5).

It is important to follow one of the standardised methodologies that are available. Without a standardised protocol, you risk making incorrect judgements about whether a bacterial isolate is sensitive or resistant and limit the comparisons you would be able to make within your own experiments.

Common problems with technique include not swabbing the whole of the plate with the inoculum and therefore not getting an even lawn of bacteria. To avoid this, make sure the plate is swabbed in three orientations to create a confluent lawn.

Moreover, if the initial inoculum is too light or heavy, it will not create a suitable lawn, giving misleading zones. To avoid this, ensure the inoculum is the correct density before starting work.

Wet plates can cause the bacterial inoculum to smudge or run across the plate, making it difficult to measure the zone sizes accurately. Ensure your plates are properly dried before using them for this technique. Antibiotic discs can fall off the plate upon inversion, leading to a missing data point for

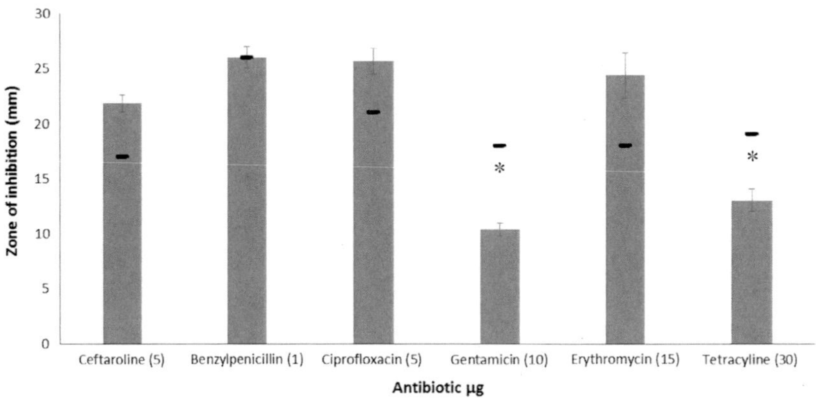

Figure 4.5 Zones of inhibition (in millimetres) generated by a range of antibiotics against one strain of *Staphylococcus aureus*. The *black horizontal lines* represent the breakpoint beneath which isolates are resistant (marked with *asterisks*). Error bars represent standard deviations of the mean.

that antibiotic. To avoid this, check that antibiotic discs are firmly in place before inverting the agar plates for incubation.

Some antibiotic discs can generate large zones of inhibition that affect the zones of inhibition generated by the disc next to them. If the zones are so large that they merge into one another, there are options for measuring the radius of the zone of inhibition using the opposite side of the zone of inhibition to the part that is affected. Ideally, however, it is best to retest the antibiotics involved on a plate on which they are more spaced out so that no merging of zones occurs.

4.1.3 Antibiotic strip tests

The Etest is another method of determining the antibiotic susceptibility of an organism used in conjunction with the agar plate lawn method. The strips themselves are thin strips impregnated with a predefined gradient of antibiotic. Using an antibiotic strip to test the sensitivity of an organism, it provides the concentration of antibiotic required to inhibit the bacteria being tested. The lowest concentration of antibiotic needed to inhibit an organism is known as the minimum inhibitory concentration (MIC) of the organism. The antibiotic strip method can be used if the MIC is needed and agar lawn plates are already being set up or if facilities to read broth dilution plates are unavailable in the laboratory. The antibiotic strips can be purchased from several manufacturers, including bioMérieux (https://www.biomerieux.co.uk) and Liofilchem (https://www.liofilchem.com/en/).

Briefly, the antibiotic disc susceptibility testing method is thus:

1. Agar lawn plates should be prepared in the same way as described in Section 4.2 for the disc diffusion method.
2. Ensure the antibiotic strips have been allowed to come up to room temperature before using them and always store unused strips according to the manufacturer's instructions. Once the bacteria have been inoculated onto the agar plate to form a lawn, the antibiotic strip should be carefully removed from the packet using sterile forceps. Ensure that the forceps are used to hold the strip only at the top edge of the strip, away from where the antibiotic is impregnated into the strip, because you do not want to damage the antibiotic gradient. Once the strip has been removed from its packaging, it can be placed in the centre of the plate if only one antibiotic is being tested, in pairs on a 9-cm agar plate, or top to tail. Up to six strips can be tested by placing them radiating out in a star shape from the centre of a large (15-cm) agar plate. The highest concentrations of antibiotic should be placed toward the edge of the plate if multiple antibiotic strips are being tested (Fig. 4.6). The strips should be placed so that the scale is facing up so it can be read.

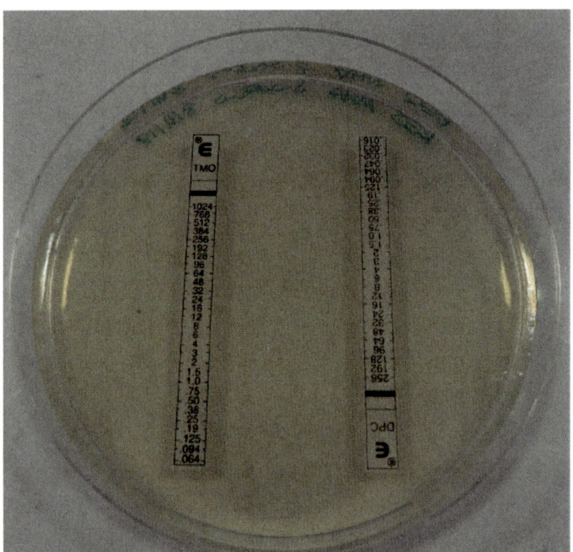

Figure 4.6 Example of how antibiotic strips can be laid out on a 9-cm agar plate to ensure effective use of space on the agar. Note that the strips are laid out top to tail. This ensures that areas where the highest concentrations of antibiotics are and where the largest zones of inhibition will occur are at opposite ends of the agar plate for each strip.

As with the discs, ensure that the whole strip is in contact with the agar. Sometimes air bubbles can become trapped under the strip. If any large bubbles are observed, try to move the bubble up and out from under the strip by gently pushing the bubble up from the low-concentration end of the strip to the high-concentration end. Small bubbles should not affect the zone of inhibition generated, so they can be left under the strip.

The plate is then incubated in an inverted position at 35°C for 18–24h, but this can change depending on which organisms you are using and which manufacturer's products are being used. Check the manufacturer's guidelines to ensure that the most reliable results are obtained. After the incubation period, remove the plate from the incubator and check that there is a confluent lawn of growth on areas of the plate where the bacteria have grown. If the growth looks wrong (too heavy or too light), disregard the plates and repeat the experiment. If the bacteria are susceptible to any of the antibiotic strips used, a zone of clearing will be seen on the agar. To read the MIC, look for the point at which the zone of inhibition intersects the antibiotic strip. The MIC is the concentration of antibiotic written on the strip at that point (Fig. 4.7). If the

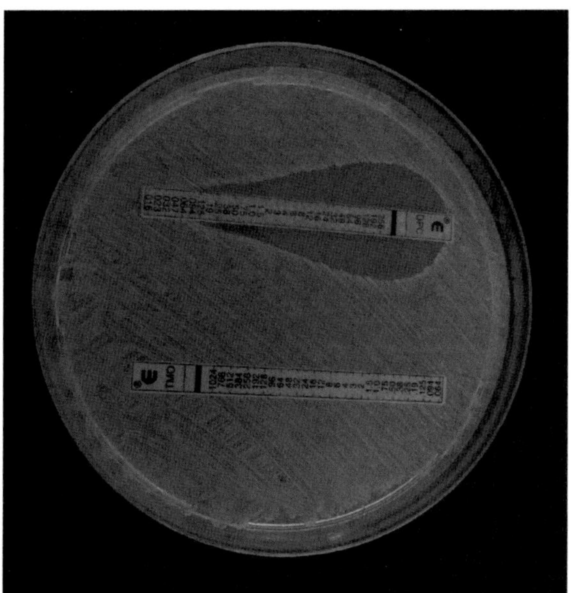

Figure 4.7 Antibiotic strip test using daptomycin (DPC) and temocillin (TMO). There is no minimum inhibitory concentration (MIC) for TMO. The point at which the zone of inhibition intersects the antibiotic strip and for DPC is at 0.125. Therefore, the MIC for this antibiotic is 0.125.

bacterial growth intersects the strip between two values, the higher value will give the more conservative MIC, which means the bacteria is less likely to be accidentally recorded as susceptible when resistant. If results fall between two concentrations, other options are to rerun the susceptibility test or confirm the MIC by another method.

4.1.4 Broth dilution

Another method that can be used to determine the sensitivity of bacteria to antibiotics is the broth dilution method. Like the Etest method, this will give you the MIC of an antibiotic required to inhibit the growth of the bacteria you are testing. In addition, this method provides you with the opportunity to test for the minimum bactericidal concentration (MBC), the lowest concentration of antibiotic needed to kill the bacteria you are testing. This is carried out in an additional step that cannot be performed from the agar plate methods described previously. For this type of test, if you want to ensure reproducible results that can be compared with other published data, or if you want to be able to check whether the bacteria are sensitive or resistant according to published breakpoints, you should follow a standardised method from an organisation such as EUCAST or CLSI. An outline example of this type of method is given here; however, you should check for details at the relevant organisation's website.

Before starting this method, ensure you have all of the equipment you need to avoid having to interrupt the experiment after you have started. You will need a bacterial culture, a Bunsen burner, 96-well microtitre plates (preferably with a lid), inoculating loops, sterile saline and sterile broth (usually Mueller–Hinton Broth), the antimicrobial agent being tested, a low-volume pipette and tips (100–200 μL) and a multichannel pipette if you have one available (Fig. 4.8).

In this method, you will need to prepare a range of antibiotic concentrations before you can start setting up the microtitre plate. To ensure you use a suitable range of antibiotic concentrations so that the test is likely to find an MIC, refer to clinical breakpoint tables to see where the resistance breakpoint is. Using a doubling dilution technique, create a range of antibiotic concentrations. Ideally, as suggested previously, these should go above and below this breakpoint to allow you to determine whether the bacteria are sensitive to the antibiotic you are testing. The number of bacterial strains being tested and the number of replicates that are undertaken will influence how you prepare the doubling dilution antibiotic range.

There are two ways in which you can create the concentration series needed. A stock of each concentration of antibiotic can be created from

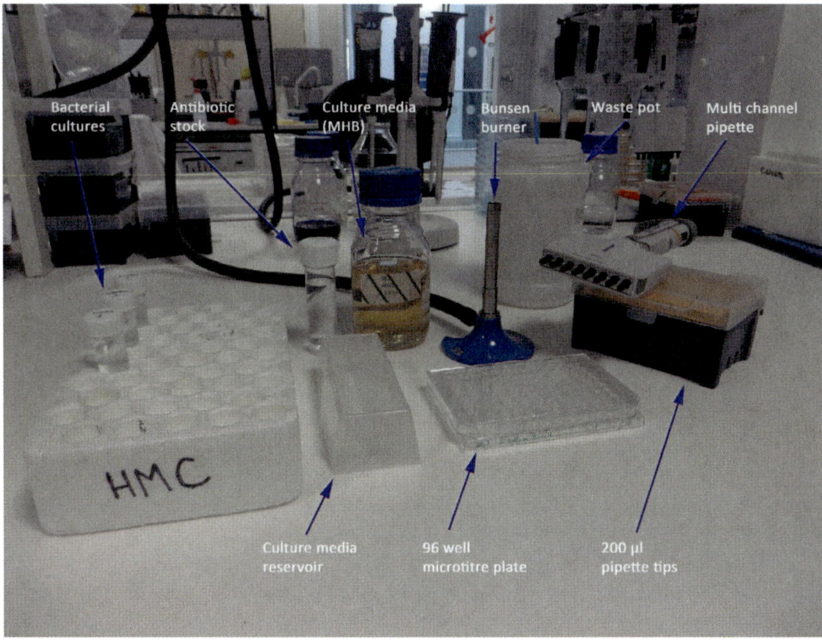

Figure 4.8 Typical setup used for minimum inhibitory concentration testing with all of the equipment needed arranged close by on the bench. *MHB*, Mueller–Hinton Broth.

an initial stock of the highest concentration needed in universal tubes or Falcon tubes. Then, this can be pipetted into the microtitre plates. Briefly, you would need to create a stock solution of antibiotic (e.g., 512 mg/L antibiotic). Pipette 3 mL of this into 3 mL sterile culture broth. Mix thoroughly (this gives you 6 mL 256-mg/L antibiotic). Pipette 3 mL of the new solution into 3 mL sterile culture broth and mix thoroughly. Keep doing this until you have created the range of antibiotic concentrations that you want to test. The amount you pipette between tubes will depend on how many wells you need to fill, which will depend on how many bacterial samples you are testing. Work out the volume of antibiotic you will need before you start this process.

A stock of the highest concentration of antibiotic can also be made in a universal or Falcon tube, but then the doubling dilutions can be carried out in the microtitre plate by filling each test well with 100 µL sterile culture media before adding 100 µL of the top concentration to the first set of wells, mix thoroughly, and then pipetting 100 µL from that well into the next and continuing in that way until you get to the final well. A total of 100 µL from the final test well will need to be pipetted into a waste bin or the volumes

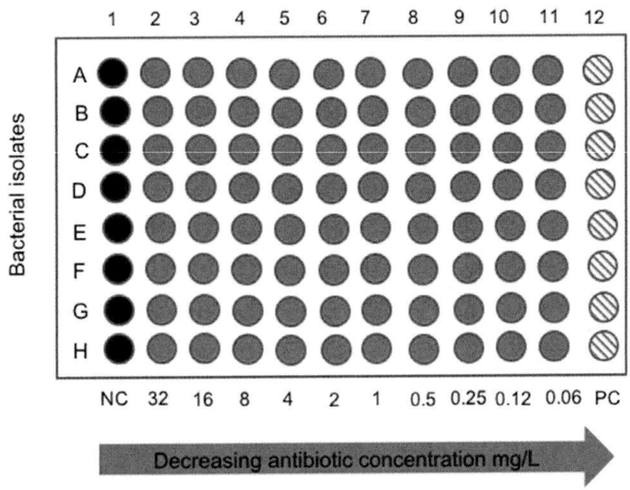

Figure 4.9 Microtitre plate displaying a horizontal layout allowing a wide range of concentrations (32–0.06 mg/L) to be tested. In this setup, if testing in triplicates, only two full sets of triplicate isolates will fit onto the plate. The third replicate of the third isolate would need to go into another plate. Alternatively, different isolates can be tested within the plate and multiple plates can be used to run the triplicates. *Black wells* represent negative controls (NC), *striped wells* represent positive controls (PC), and *grey wells* represent test wells. Each column contains a different concentration of antibiotic.

in the plate will be wrong. Make sure that none of the antibiotic is pipetted into either the positive or negative control wells; only the test wells should contain the test antibiotic.

For example, when testing the sensitivity of *Staphylococcus aureus* to linezolid in which an MIC of ≥4 mg/L is resistant and an MIC of <4 mg/L is sensitive, there is a choice of two antibiotic concentration ranges that could be prepared in the plate, depending on how the plate is orientated and how many isolates need to be tested. In the first format, in which the antibiotic gradient runs horizontally across the plate, a negative and positive control as well as a range of 32–0.06 mg/L would be possible (Fig. 4.9).

If the gradient runs vertically, a negative and positive control as well as a range of 16–0.5 mg/L would be possible (Fig. 4.10). Which format you choose to use will depend on how many isolates you are testing.

Whichever way the dilutions are carried out, the final volume of antibiotic in each well before the bacteria are added should be 100 µL. When the antibiotic concentration range has been set up in the plate, the bacteria can be added. Moreover, once 100 µL of bacteria is added to each well, the concentration of antibiotic in each well will halve.

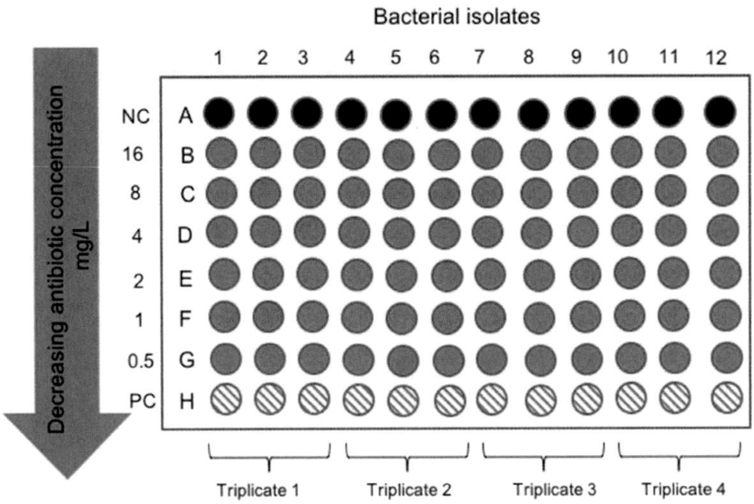

Figure 4.10 Microtitre plate displaying a vertical concentration layout that has a narrower range of antibiotic concentrations (16–0.5 mg/L). This setup allows complete sets of triplicate isolates to be tested in the same plate. *Black wells* represent negative controls (NC), *striped wells* represent positive controls (PC), and *grey wells* represent test wells. Each row contains a different concentration of antibiotic.

Once the control wells have been prepared, the test wells will need to have the bacterial inoculum added. Start by aseptically picking a few bacterial colonies from an overnight agar plate culture and suspend these in sterile saline to give a density equivalent to a 0.5 McFarland standard. Depending on which bacteria you are testing, there may be a further dilution step at this point, so check standardised methods for guidance. Once appropriately diluted, 100 μL of the culture will need to be pipetted into each well, including the positive control wells but excluding the negative ones.

When the plate is set up, there should be 100 μL of antibiotic (varying concentrations) plus 100 μL bacterial cells in each test well. The positive control wells will contain 100 μL of broth plus 100 μL bacterial cells and no antibiotics. The negative control wells will contain 200 μL broth with no antibiotics and no bacteria.

When preparing these plates, ensure that all the work is done aseptically and that the correct volume and concentration of antibiotic go into each well. All wells should contain 200 μL when the plate is completed. If using a multiwell pipette, check when drawing up liquid that the same amount is in each pipette tip. If you bend the tips while drawing up liquid or push the tip too hard against the slide of the container, the pipette may not draw up the right amount of liquid in each tip. It is good practice to note in your laboratory notebook what is in each well of the plate and anything that happens during the preparation which could affect the results (e.g., if you think you might have pipetted twice into one well, missed a well or accidently contaminated something). This will allow you to pick up on any errors or issues more quickly when observing the results.

Once the plates are completed, they will need to be read in a spectrophotometer to provide a 0-h reading. Place the microtitre plate into the spectrophotometer and ensure it is in the right orientation. Set the spectrophotometer to the wavelength needed: for example, 600 nm. With many software packages, there is an option to label the wells on-screen to match those in the plate. Once the wavelength is set and the template has been set to measure the optical density (OD) in all of the wells that are required, read the plate. Make sure the data generated are saved under a name that can be found again. The spectrophotometer might be used by many people or you might be reading multiple plates. If the reading is saved as something such as the date, it might be difficult to find again when it is looked for it among all of the readings performed in a day. Ensure that there is enough information to be able to identify each reading back to the original plate. Most spectrophotometers will allow data generated to be saved or exported into a programme such as Excel. This will make it easier to analyse the data at a computer not connected to the spectrophotometer.

Once the 0-h reading has been completed and saved, incubate the plates at approximately 35°C for 16–20 h. Once the incubation period is over, look at the plate by eye to pick up obvious problems or contamination issues and then read it again at the same wavelength at which the 0-h reading was carried out and save the data. The data are then ready to be analysed.

It is good practice to run these tests using three biological replicates and three technical replicates for each of biological replicate (this will generate nine data points for each well tested). Ensure that a control organism with known sensitivities is used to verify that the results are reliable and not caused by an issues with the method or media.

Once the data have been exported from the spectrophotometer, they will need to be analysed and presented in a way that is easy to read, such as a line graph. It is usually a good idea to subtract the 0-h and negative control data from the 20-h test data. If each test well has been done in triplicate or multiples, the mean value will need to be calculated. For example, when working with tetracycline, the values for each concentration would be added together and divided by the number of replicates. This will generate one mean value for each concentration of tetracycline. Once there is a mean value for each concentration, those figures can be plotted onto a graph. The concentration at which there is no turbidity (growth) is the MIC. The standard deviation

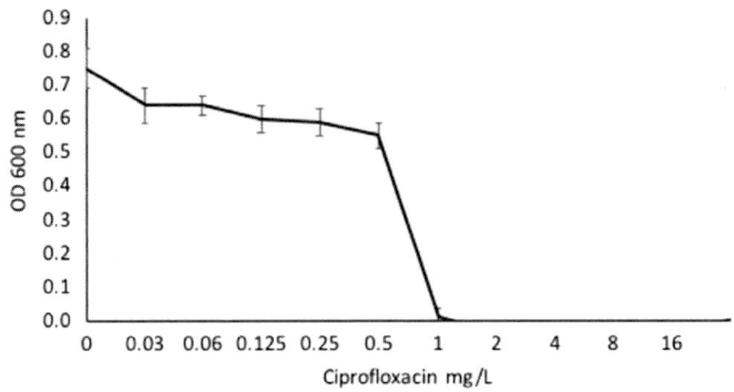

Figure 4.11 Example of how to present minimum inhibitory concentration (MIC) data from a broth dilution experiment. The MIC (1 mg/L) of ciprofloxacin is shown when tested against an isolate of *Staphylococcus aureus*. Error bars represent the standard deviation of the mean. *OD*, optical density.

should also be calculated for each concentration to ensure that replicates are similar and can be added to the graph as error bars (Fig. 4.11).

The standard deviation can be calculated in a programme such as Excel by selecting an empty cell, pressing Equals, selecting standard deviation from the Functions option bar (often at the top left in Excel and displayed as STDEV or similar). Then, select all of the replicates for the concentration of interest and press Return. This process can be repeated for each concentration. Once the standard deviations have been generated and plotted onto the graph, they should provide an easy way to seeing whether your replicates were similar or whether there is a lot of variation between samples. A large range of variation with samples could indicate that the experiment needs to be repeated. For some antibiotics, there might not be complete reduction of OD to 0. If there is no MIC for the antibiotic tested, it is possible to present the lethal dose (LD) as 50 or 90, which is the amount of antibiotic needed to kill 50% or 90% of cells and is often represented as LD50 or LD90. It can be calculated for any proportion of cell death, but 50 and 90 are commonly used.

The MBC can now be determined from the original MIC plate. After reading the plate in the spectrophotometer, pipette 10 μL from every well that has no growth in it onto an agar plate. Wait for these spots of liquid to dry in the plate, and then incubate the agar plate at 35–37°C overnight. After incubation, inspect the plate for growth. The lowest concentration where there is no bacterial growth is the MBC. The antibiotic is normally considered to be bactericidal if the MBC is not more than four times the MIC. If the MBC is more than four times the MIC, the antibiotic is usually considered to be bacteriostatic.

The MIC can also be determined using an agar dilution method. This is similar to the broth dilution method in that an antibiotic would be added to agar in a range of concentrations before the agar has set. This can be carried out in a Petri dish. The size of the dish does not matter, but the volume required to fill it is important. The volume of antibiotic that needs to be added to the agar has to be calculated based on the volume of agar used to create the agar plate. Ensure that the agar is not too hot when adding the antibiotic. If possible, wait until the agar has cooled to at least 50°C before adding the antibiotic. To ensure that there is an even distribution of antibiotic throughout the agar, swirl the Petri dish once the agar has been added. When swirling the agar in the Petri dish, make sure the agar does not spill out over the side of the plate or up into the lid. Leave the agar to cool and set. Once the agar is set, a known quantity of bacteria can be spotted on the surface. Adding a standard inoculum to each agar plate allows reproducibility between tests. After the bacteria have been added and have dried in the plate, incubate the inverted plates at 16–20 h at about 35°C. After incubation, inspect the plates for growth. The minimum concentration will be the plate containing the lowest concentration of antibiotic that displays no growth. This method can be useful if you have a series of bacterial isolates to test, because they can be spotted onto different areas of the same agar plate. Unlike the broth dilution, it is impossible to determine an MBC from these plates.

4.1.5 Synergy

The use of combined antibiotics to enhance the activity of one or all of antibiotics that have been combined is an approach that has become popular since increasing antibiotic resistance has been observed. Combining antibiotics to treat infection can reduce the amount of each antibiotic needed, reducing associated host toxicity and potentially leading to quicker eradication of infection. Whether a particular antibiotic will improve the activity of another one will depend on the antibiotics being combined and the bacteria being tested. Being able to identify correctly which antibiotic combinations are likely to improve efficacy is important. Some ways in which you can check antibiotic interactions are described next.

The methods described in Sections 4.2–4.4 can be adapted to determine whether there are interactions among the antibiotics being tested. Interactions among antibiotics or antibiotics and other compounds are important because they can highlight where compounds inhibit one another or where they increase the efficacy of a compound beyond what you would expect when they are used alone. The antibiotics could have different interactions, including;

Antagonism: when the combined effect of two or more antimicrobial agents reduces the efficacy of those antimicrobial agents compared with their use alone;

Synergy: when the combined activity of two or more antimicrobial agents is greater than the expected activity of the antimicrobial agents when added together; and

No interaction: when the combined antibiotics cause no increase or decrease in the efficacy of the other antibiotic.

Disc diffusion is the first method which can be used to determine whether there are interactions between antibiotics. The disc tests can identify both antagonistic and synergistic interactions between antibiotics. The plates should be prepared as described in Section 4.2 and the discs of interest should be placed next to each other on the plate, approximately 20 mm apart. Incubate the plate as outlined in Section 4.2. When examining the plates the following day, if antagonism is present, the zone of inhibition between the discs will display a blunted shape of one or both of the zones involved (Fig. 4.12). This indicates that the antibiotic that shows a blunted zone of inhibition is no longer as effective as it was when used on its own.

If synergy occurs between antibiotic discs, there will be an increase in the zone of inhibition on the side of the zone nearest the antibiotic disc with which it is interacting. This interaction indicates improved activity of the antibiotic when it is used in conjunction with the second antibiotic (Fig. 4.13).

To looks for interactions using a broth method, you will need the same equipment as for the MIC broth dilution, so ensure that you have a bacterial culture, a Bunsen burner, a 96-well microtitre plates (preferably with a lid),

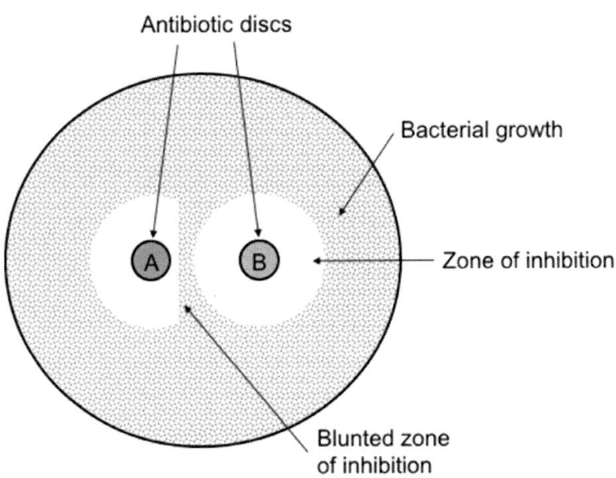

Figure 4.12 An antagonistic interaction is seen between two antibiotics (A and B) in this diagram of antibiotic disc testing. Blunting in the zone of inhibition for Antibiotic A is seen because interaction with Antibiotic B has reduced the efficacy of Antibiotic A compared with Antibiotic A used alone.

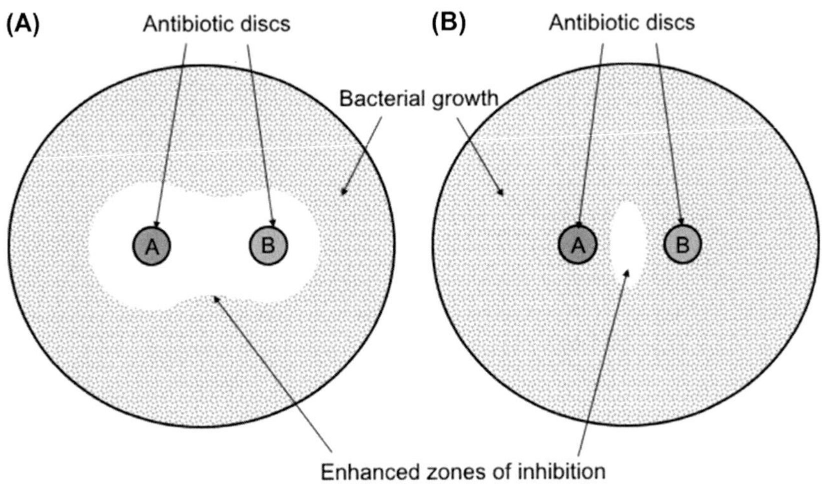

Figure 4.13 (A and B) Synergistic interactions. Increased zones of inhibition are seen in (A), where the two antibiotics interact, indicating that this combination of antibiotics is synergistic. (B) A zone of inhibition has appeared in the area between the two discs, indicating a synergistic interaction. The antibiotic discs do not inhibit the bacteria when used alone, because no zone of inhibition is seen around the rest of the disc.

inoculating loops, sterile saline and sterile broth (usually Mueller–Hinton Broth), the antimicrobial agents being tested, a low-volume pipette and tips (100–200 μL) and a multichannel pipette if you have one available.

Before working out whether there are interactions between antibiotics, an MIC will need to be established for each antibiotic alone. This can be done in the plate in which the synergy experiment is being carried out, because there will be an antibiotic-alone series of test wells for each antibiotic. However, it is probably better to establish the antibiotic MIC using a standard broth dilution experiment as described earlier, so that when preparing the synergy plate, you have already established the range needed to give several concentrations of antibiotic both above and below the original MIC.

Once the series of antibiotic concentrations has been prepared, it will need to be pipetted into the 96-well plates. Pipette 50 μL of each antibiotic into the wells. You can set up the plate as shown in the template in Fig. 4.14. Once the antibiotics have been added, 100 μL of bacteria (prepared as described in Section 4.4) can be pipetted into all of the test wells and positive control wells. Do not add bacteria to the negative controls.

The concentration of antibiotics going into the well will halve in wells where you are adding both antibiotics, and it will halve again when the bacteria are added. For example, when adding 50 μL of Antibiotic

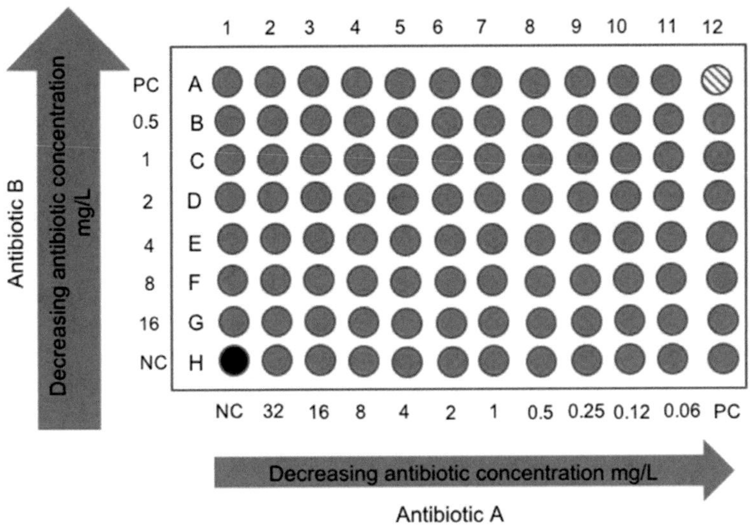

Figure 4.14 Sample plate layout that can be used for checkerboard testing of antibiotics (it can be adapted to test other antimicrobial agents). *Black well* represents the negative control (NC) well, *striped well* represents the positive control (PC) well and *grey wells* represent the test wells. Each grey well will contain different combinations of concentrations of the two antibiotics being tested.

A at 256 mg/L to 50 μL Antibiotic B at 256 mg/L, there will be 100 μL of Antibiotic A and B at 128 mg/L. When 100 μL bacteria is added, the well will contain 200 μL Antibiotic A and B at 64 mg/L. When marking the plate or writing in a laboratory notebook, make it clear whether the plate layout is describing the original concentration created in the dilution series or the final concentration in the well once the antibiotic and bacteria have been added.

The plate can now be read in the spectrophotometer at 0 h. Incubate at 35°C for 16–20 h and then read it in the spectrophotometer again. Analysis of these data is similar to that carried out on the microbroth dilution plate. Subtract the 0-hand negative control values from the 20-h values and work out what the new MIC is for each antibiotic. The new MIC for each antibiotic can be calculated for every well in which the second antibiotic is at a subinhibitory concentration. The fractional inhibitory concentration index (FICI), which indicates which type of interaction is occurring, can be calculated using the original and new MIC values for both antibiotics by using the calculation:

$$\mathrm{FICI} = \mathrm{FIC}\,(\mathrm{A}) + \mathrm{FIC}\,(\mathrm{B})$$

where FIC(A) = MIC Antibiotic (A) in combination/MIC Antibiotic (A) alone; and FIC(B) = MIC Antibiotic (B) in combination/MIC Antibiotic (B) alone.

For example, if:

$$\text{FIC}(\text{A}) = 0.5/16 = 0.06$$

$$\text{FIC}(\text{B}) = 0.5/2 = 0.25$$

$$\text{FICI} = 0.06 + 0.25 = 0.3$$

In this example, the FICI value indicates a synergistic interaction between Antibiotics A and B. There are different cutoffs depending on which report you read, but values can be interpreted using Table 4.3, which gives FICI interpretations as suggested by Odds (2003). There are other interpretations of FICI values provided in the literature, such as those suggested by EUCAST (2000); however, the values given here allow for experimental variation.

4.1.6 Novel compound testing

To test novel or non-antibiotic antimicrobial agents that are not included in standardised methods but that can be dissolved in broth or phosphate-buffered solution, a couple of options are available to determine MICs, adapted from standardised methods described in Sections 4.2–4.4.

Blank paper discs are available for purchase from suppliers such as Sigma-Aldrich and Thermo Fisher Scientific. These can be used in a manner similar to the antibiotic disc testing method. Pipette a known concentration and amount of test material onto a blank paper disc and allow the liquid to soak into the disc. The disc can then be used in the same manner as the antibiotic discs described in Section 4.2. Keep the other parameters the same, including the bacterial inoculum, the temperature and time of incubation and the measurement of the zones of inhibition. Using the blank discs in this way will allow you to create reproducible zones of

Table 4.3 Interpretation of fractional inhibitory concentration index values, taking conservative approach to allow for testing variation suggested by Odds (2003).

Interaction	Value
Synergy	≤0.5
Indifference	0.5 to ≤4
Antagonism	>4

inhibition from the discs, and therefore comparable results. This should allow you to assess whether the novel compound can inhibit bacteria and also to compare effectiveness between different strains or species of bacteria. There are unlikely to be breakpoints available for a novel compound, but if the concentration and volume of compound are kept consistent, this adapted method should provide useful information about the activity of the compound.

If the compound is soluble, it is also possible to adapt the microbroth dilution test described in Section 4.3. Replace the 100 μL of antibiotic with 100 μL of the antimicrobial compound (at a range of concentrations) and keep all other parameters the same, including the bacterial inoculum, incubation time and temperature and data analysis. The test for an MBC can also be used in the same way as described in Section 4.3. It is unlikely that there will be breakpoints available for novel antimicrobial agents, but adapting standard tests can provide a reproducible way to establish both an MIC and MBC.

When working with a material such as an antimicrobial wound dressing or antimicrobial surface, it may not be possible to test using these techniques. However, several variations on standard antibiotic testing can be used.

It is possible to assess the antimicrobial effect of exposure to a material or surface on test bacteria as calculated compared with a control. For example, if testing a new dressing impregnated with an antimicrobial agent, a protocol such as the one suggested here could be followed.

If testing one type of bacteria against one type of dressing, start with multiple pieces of dressing all cut to the same size (e.g., 2 cm^2). To test on agar, prepare a bacterial lawn plate as if preparing for a disc diffusion test (Section 4.3). Place the material or dressing directly onto the plate, incubate the plate at about 35°C for 16–20 h and measure the zone of inhibition with callipers as in Section 4.3. Like the blank disc method, this method will allow you to assess whether the novel compound can inhibit bacteria and also allow comparison of its effectiveness between different strains or species of bacteria. If the material or dressing is impregnated with an antimicrobial agent or has an antimicrobial coating, it is a good idea to run a test sample that does not have the coating, to see whether the material alone has a antimicrobial effect.

To test the novel material or dressing directly, create a bacterial standard with a known number of cells (such as the McFarland standard). Inoculate each piece of dressing with a set volume of the bacteria (e.g., 20 μL). Each piece of dressing can be placed in a sterile Petri dish or universal container.

Incubate the material with the bacteria for a set period. This time could be set to match the standardised testing or adapted to give a customised set of results. If possible, and if testing a completely novel material, it may be useful to inoculate enough pieces of test material for a time course. This would allow testing to occur at multiple time points (e.g., 1, 2, 4, 8, 12, 18 and 24 h).

At the given time point, place the test material into 5 mL of one quarter-strength Ringer's solution and vortex for 2 min to release any bacteria. Create a dilution series from this solution using the Miles and Misra technique outlined in Chapter 2, Section 2.5 and pipette 50 µL onto an agar plate. Incubate the inverted agar plate overnight at 35–37°C, count the colonies the following day and calculate the colony forming units per millilitre as outlined in Chapter 2, Section 2.5. This should allow you to see whether the material has an antibacterial effect and how long it takes to cause that effect.

As with all experiments, it is essential to run control samples alongside the test samples to ensure the reliability of results. If possible, when testing a novel dressing or material with an antimicrobial impregnated or coated onto the material, run a uncoated version of the material with the bacteria as a positive control. This should allow you to check whether the material itself has an impact on the bacteria being tested. Running a negative control with no bacteria added allows any contamination of the experiment to be identified.

This type of experiment could also be run by adding the novel material or dressing to a broth culture of bacteria of a known starting density and vortexing the broth and dressing together at the various time points, including a 0-h cell count. An aliquot from this broth can then be processed at each time point using the Miles and Misra technique. If testing the material in broth, a positive control using untreated or coated material can be run. A bacteria-only control (no material or dressing) could also be run to determine whether the dressing alone has an impact on the bacterial population. These types of methods can be adapted to suit the different types of novel antimicrobial agents (compounds, materials and surfaces) being tested. If you have to adapt away from standard methodologies, it is best to make sure that positive and negative controls are run whenever possible and to minimise variables. This may involve optimising the method, but it will allow the maximum amount of useful information to be gathered from testing.

Because of increasing antibiotic resistance seen in a wide range of bacteria, the ability to test bacteria for susceptibility to both conventional and novel antimicrobial agents is a fundamental skill needed by many microbiologists. This chapter has provided an overview of different methods by which this testing can be carried out and has described ways in which the data generated can be analysed and interpreted.

4.2 Notes page

Record observations and notes here.

CHAPTER FIVE

Cell culture-based infection models

Contents

5.1 Introduction

It is possible to glean fundamental information regarding microbial behaviour using analysis of growth patterns, microscopy and antimicrobial testing. To begin to understand how these observations better relate to the infection process, simple in vitro infections models can be used that rely on cell culture. These allow the researcher to ascertain how quickly bacteria will infect a relevant cell line (e.g. representing the skin or gastrointestinal epithelia) and what impact this has on the infected cells in terms of their immunological response and survival. This chapter will help you to decide how best to use cell culture-based infection models for your studies and what the most relevant parameters for study might be.

5.1.1 Which species of bacteria are you working with and which is the best cell line to use?

Are you working with a known pathogen or a commensal organism? In the first instance, the answer will help you to decide whether a cell culture-based model is appropriate for your studies. If you are studying a pathogenic bacterium, a cell culture-based infection model has the potential to provide you with salient information, such as how quickly the organism infects and/or kills a cell line, how readily it attaches to the surface of the cells and whether it is internalised

Bacteriology Methods for the Study of Infectious Diseases
ISBN 978-0-12-815222-5
https://doi.org/10.1016/B978-0-12-815222-5.00005-5

or invades the cell line. If you are using a commensal organism or opportunistic pathogen, a cell culture-based model can still be used to investigate how the bacterium interacts with the cell line in terms of attachment, and the type of response it elicits in the cell line: for example, the cytokine profile. This has the potential to help you to understand the role a commensal might have in maintaining immune homeostasis and how bacteria can survive in the host environment; opportunistic pathogens could identify what elicits pathogenicity. By considering these factors, you are essentially coining your research question.

For cell culture-based infection models to be relevant, you need to identify the typical colonisation or infection sites for the bacterium you are studying. For example, this might be the respiratory tract, gastrointestinal tract, skin, or genitourinary tract. This will determine the type of cell line you choose; it is important to ensure that you choose a physiologically relevant cell line to work with (this is true for humans and mammalian infections). Some examples of human cell lines available from the American Type Culture Collection (ATCC) and their associated physiological site are given in Table 5.1.

There are likely to be multiple cell lines applicable to your work, so it is important to research each one thoroughly before deciding which is best to use. One important consideration is whether to employ primary or immortalised cell lines. Primary cell lines are derived directly from living tissue (for example, via biopsy) and so have undergone only a small number of passages (being split [see Section 5.2]), which makes them a more representative

Table 5.1 Examples of immortalised cell lines that correspond to different anatomical sites in the human body, and which are part of the American Type Culture Collection (ATCC).

Tissue	Morphology	Properties	ATCC identifier
Bone	Mixed	Adherent	CRL-8303
Colon	Enterocyte	Adherent	CRL-2102
Epidermis	Epithelial	Adherent	CRL-1555
Eye/cornea	Epithelial	Adherent	CRL-11515
Eye/lens	Epithelial	Adherent	CRL-11421
Kidney	Epithelial	Adherent	CRL-11268
Lung	Epithelial	Adherent	HTB-53
Lymphocyte	Lymphoblast	Suspension	CRL-2570
Pancreas	Epithelial	Adherent	CRL-1739
Pharynx	Epithelial-like	Adherent	CRL-3212
Retroperitoneum	Fibroblast-like	Adherent	CRL-3044
Skin	Fibroblast	Adherent	CRL-1533
Spleen	Lymphoblast	Suspension	CRL-9300
Stomach	Epithelial	Adherent	CRL-1739

model of in vivo conditions. Typically, once purchased, these are cultured, and stocks are prepared for storage at −80°C. Fresh cells must be cultured for each experiment and have a finite life span. Primary cells have not been modified in any way but their morphology can change in culture as they age, which is why it is important to use them fresh each time and work with early passages. Because they have undergone no modifications, primary cells are generally less hardy than immortalised cell lines, which can make them more difficult to culture, often requiring specialised media and supplements. Immortalised cell lines either are derived from tumours or have been artificially modified and so proliferate indefinitely. Advantages of using immortalised cell lines are that they are easier to culture than primary cell lines, are well-characterised, and are genetically identical and homogeneous. A drawback to such models is that immortalised cell lines are not normal cells, because they have been modified in some way to make them immortal; consequently, any biological effects observed may not be truly representative of the in vivo state.

Table 5.2 lists some commonly used primary cell lines.

As briefly mentioned, primary cells can be more difficult to culture than immortalised cell lines, and each one typically requires a different growth medium and, in some cases, supplements. Once you have chosen the cell line appropriate for your work, it is important to identify the media needed to culture the cells and the appropriate growth conditions. For example, some cell lines must be grown in incubators with a 5% CO_2 atmosphere. Some cell lines will differentiate or can be induced to differentiate. For instance, over time, colon cell lines can become villiated. Check for these characteristics if they are necessary to your study. Tables 5.1 and 5.2 indicate whether the cell lines are adherent or grow in suspension. Consider the infection you are trying to model when choosing the property of the cell

Table 5.2 Examples of primary cell lines that correspond to a variety of different anatomical sites in the human body, and which are part of the American Type Culture Collection (ATCC).

Tissue	Morphology	Properties	ATCC identifier
Aorta	Endothelial	Adherent	PCS-100-011
Bladder	Endothelial	Adherent	PCS-420-010
Bone	Spherical	Suspension	PCS-800-012
Eye/cornea	Epithelial	Adherent	PCS-700-010
Respiratory tract	Epithelial	Adherent	PCS-300-011
Respiratory tract	Fibroblast	Adherent	PCS-201-020
Skin	Fibroblast	Adherent	PCS-201-012
Vagina	Epithelial	Adherent	PCs-480-010

line. For studies that investigate attachment and invasion of bacteria into a given cell line, an adherent property is far easier to work with (see Section 5.2). Finally, cell lines are classified according to Advisory Committee on Dangerous Pathogens (ACDP) regulations, much like bacteria; therefore, it is imperative to find out what the ACDP rating is for the cell line you choose to work with, to ensure that you have the right laboratory facilities in which to work. Information specific to each cell line can be found in the ATCC database (https://www.lgcstandards-atcc.org/).

5.1.2 Cell culture and enumeration

Like bacterial culture, cell culture must be undertaken aseptically. Often, laboratories will have dedicated suites for this. Work is ordinarily carried out in a biosafety cabinet, but this does not mean that you can be complacent when handling cells, media and consumables. Before beginning work, locate all of the items you need, including tissue culture flasks, media and supplements, pipette tips, serological pipettes, a waste container filled with disinfectant, disposable gloves, 70% ethanol in a spray bottle and a haemocytometer for enumeration. Wearing disposable gloves, spray the inside of the biological safety cabinet with 70% ethanol and wipe the surfaces. Place all of the items you need to work with into the biological safety cabinet (Fig. 5.1A). For any non-disposable items (e.g. automated pipette gun, pipettes, racks), spray the surface with 70% ethanol and wipe them before introducing them into the cabinet. Place a waste container filled with disinfectant into the cabinet. This is where you will dispose of spent media. Sterile media and supplements will need to be warmed to 37°C, so ensure these are placed into a water bath at least 30 min before use. If you are only using small quantities of a particular medium or supplement, it advisable to aliquot them into smaller portions before use, to minimise the risk for contamination.

Most cells are culture in tissue culture flasks (Fig. 5.1B), which come in varying sizes to accommodate varying volumes of media. When seeding a flask using a new or frozen stock aliquot of cells, it is best to use the smallest flask available and to seed with the entire contents of the vial.

It is necessary to check cultured cells daily by inverted microscope to assess confluence when the cells need to be passaged or when they are ready to use for an experiment. Generally, adherent cells should be seen to be immobile (i.e. attached to the bottom of the flask [a monolayer]). If large clumps appear to be floating or the monolayer is detaching from the bottom of the flask, the cells are overconfluent and need to be discarded. Cells

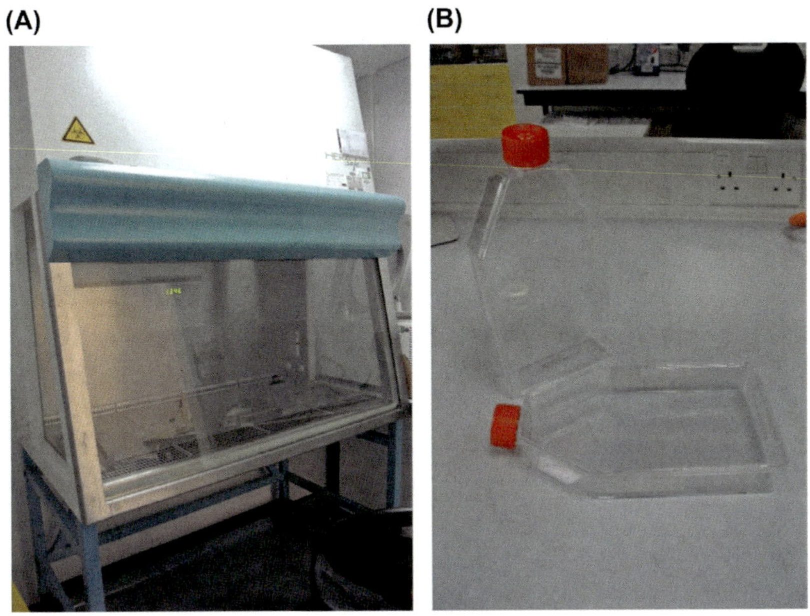

Figure 5.1 (A) Typical biological safety cabinet that can be used for tissue culture. (B) Tissue culture flasks come in a range of sizes. The largest (1 L) are shown here. When filling the flask, it must be in the upright position, but once cells are introduced, the flask must be kept flat to ensure the media cover the cells and that they are in contact with and adhere to a surface.

in suspension should appear typically spherical in shape and should refract light. No clumps should be visible; if present, the cells should be discarded.

Cells reach confluence at different rates. Information regarding typical passage times can be found when purchasing a cell line. Ordinarily, once cells are between 70% and 80% confluent (covering 70–80% of the surface of the flask), they should be split, meaning that they should be subcultured into fresh flasks or into appropriate vessels for a given experiment. Different cell lines have different recommended numbers for seeding, and it is necessary to note this information when receiving new cells. When splitting cells, you will need to have new flask, fresh media and supplements (both prewarmed) ready. Adherent cells will need to be detached from the bottom of the flask; some cells are more adherent than others. There are three different approaches you can take to achieve this. The first is to tap the side of the culture flask gently until the monolayer detaches. It will be possible to see this by eye. For more adherent cells, it might be necessary to use a cell scraper to scrape the cells physically from the bottom of the flask. To do this, discard the medium by gently pouring it away, add enough phosphate-buffered saline (PBS) to cover the monolayer and gently scrape the cells away from the surface. As soon as the cells are detached, add a small amount of medium (10% of the original volume) and resuspend them by gently pipetting. Trypsin can also be used to detach cell monolayers. In this case, discard the medium from the cells as described and wash by rinsing the monolayer in PBS sufficient to cover the base of the flask. Add a sterile solution of trypsin ethylenediaminetetraacetic acid (this should be purchased along with other cell culture-certified consumables) sufficient to cover the monolayer and place the flask into a 37°C incubator. Check the flask every 60–90 s by gently tapping the side until all of the monolayer has detached. Working in the safety cabinet, immediately add a small amount of medium and gently resuspend the cells.

Before seeding a fresh flask, it is necessary to enumerate the cells so that you know what volume to add to achieve the correct number of cells. This can be done using a haemocytometer. Ensuring that the haemocytometer and coverslip are clean (use 70% ethanol to wash) and dry, place the coverslip over the chamber and fill it with 10 μL of suspended cells, taking care not to overfill it. Place the haemocytometer onto an inverted microscope and use the ×10 objective lens with phase contrast to observe the cells. Count the cells that fall into the central grid and repeat the procedure a further two times (Fig. 5.2). To calculate cells per millilitre, multiply this average by 1×10^4. Once you know how many cells per millilitre there are, you can calculate the volume needed to seed a fresh flask of media adequately. Top up the remaining volume with appropriate media and supplements and place the flask in the incubator to allow the cells to grow, checking daily as previously described.

If the cell line you are using is slow-growing or you have cells that differentiate over time, you might need to change the media periodically without splitting the cells. To do this, remove the spent medium from the flask by gently pouring it away, and replace with fresh medium and supplements. For each cell line, it should be possible to devise a schedule for splitting based on experience of how quickly the cells grow. If you a splitting a flask and using some of the cells for an experiment (for example, you might be seeding a 96- or 24-well microtitre plate for an assay), you will have to calculate the number of cells required to seed the plates based on the volume of each well and use the data obtained from the haemocytometer cell counts.

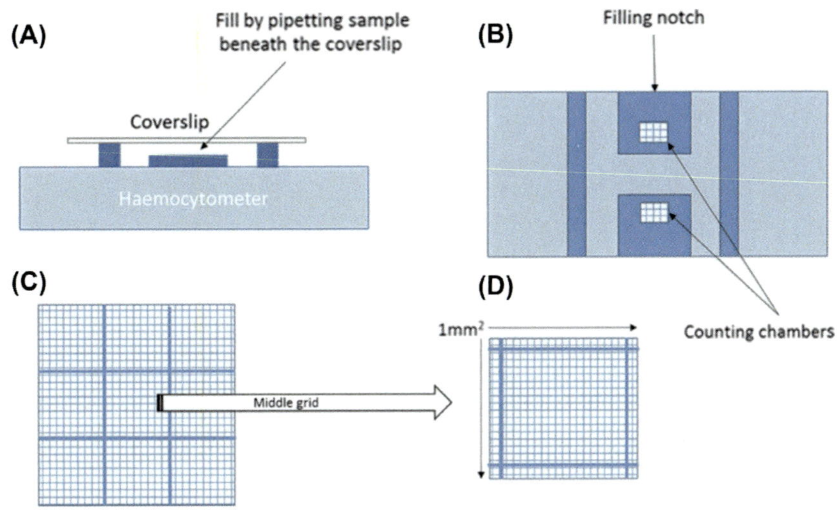

Figure 5.2 Schematic representation of haemocytometer. (A) Cross-section. (B) Top-down view. The sample is pipetting under the coverslip via the filling notch. Using an inverted microscope, observe the counting chambers that have a grid on them (C). Count the cells in the middle grid of the chamber (D). This grid has an area of 1 mm^2.

5.1.3 Viability and multiplicity of infection

Cell lines that are inoculated with bacteria will eventually die as the bacteria infect the cells and/or use the media and overrun the culture vessel. This can pose problems when trying to answer questions about the infection process and/or cell responses to infection using cell culture-based models. Therefore, it is important to establish initially what dose of bacteria can be used to infect a given cell line and for how long before the cell line succumbs. This is done by assessing the viability of the infected cells in response to dose and over a defined time period. A number of dyes are available that can be used to determine cell viability as a measure of cell metabolism. One commonly used system is the Cell Titre Blue reagent (Promega Corporation). This employs a redox dye (resazurin) and measures the ability of viable cells to convert it to the fluorescent end product, resorufin. Viable and therefore metabolically active cells are able to convert resazurin to resorufin, but nonviable cells cannot.

Typically, assays that test for viability after bacterial infection, using this or other similar dyes, require the cell line to be cultured in a 96-well microtitre plate. Different doses of bacteria are added to each well and allowed to remain in contact with the cell monolayer for a series of set time points (24, 48 and 72 h, for example). After that time, the dye is added and fluorescence is measured using a spectrophotometer. The greater the fluorescence, the more viable cells there are in the test well. It is important for this type of assay to include an untreated control so that you know how much fluorescence a healthy cell monolayer produces.

To prepare a series of different doses of bacteria for a viability assay, a serial dilution of a bacterial culture of known concentration must be prepared. To begin, grow a culture of 5–10 mL of the bacteria in question for 16–24 h. Prepare a series of 10-fold dilution ranging from 10^{-1} to 10^{-9} by dilution 10 μL of the culture in 90 μL PBS, and then dilute 10 μL of the first dilution in another 10 μL until the full series is established (Fig. 5.3A). Using aseptic technique, pipette 100 μL of each dilution into the centre of separate agar plates. Using a disposable spreader, prepare a lawn of bacteria (Fig. 5.3B) by covering the plate with the culture.

Incubate the agar plates at 37°C overnight. You should see a decrease in the number of colonies on the agar plates as the dilution increases. Count the colonies and use this formula to calculate the colony forming units (CFU) per millilitre:

$$(\text{Number of colonies counted} \times \text{dilution factor}) \times 10 = \text{colony forming units (CFU) per mL}$$

$$\text{Example}: \left(15 \text{ colonies at} 10^{-6} \text{dilution} \times \left(1 \times 10^{-6}\right)\right) \times 10 = 1.5 \times 10^{8} \text{CFU/mL}$$

Once you know the number of bacteria in each dilution, you can work out how many you will be adding to each monolayer in a 96-well plate if you add 10 μL of bacterial culture.

Add each dose and leave wait for the allocated length of time. If you have three time points, you will need three samples per dose and will have to test them in triplicate, so ensure you have enough cell monolayers to infect. After the allocated length of time, the viability assay is undertaken as described earlier.

The results of the viability assay will guide your experimental approach. If you want to study how quickly the cell monolayer is killed after infection, you might choose a dose you know will kill the cells, but over a longer time frame that you can study: for example, a total loss of viability in 72 h. If you want to understand what happened during infection over a shorter time frame, you might use a dose that will kill the samples in 24 h, but investigate what happens to the cells at 1–8 h after infection. Alternatively, you might want to establish what happens to cells that are exposed to a low but

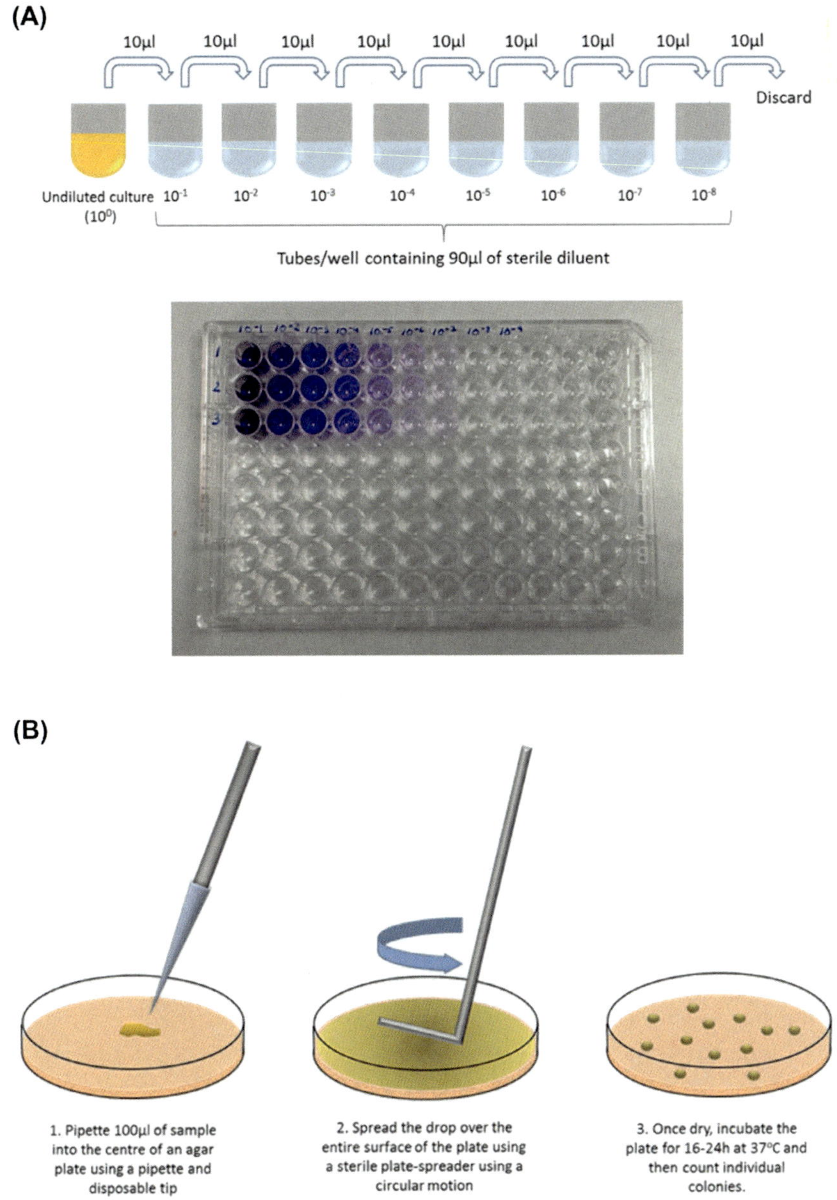

Figure 5.3 (A) Example of 10-fold serial dilution, using coloured dye to visualise the effects of diluting a sample in this manner. (B) Preparing a lawn plate of the diluted samples using a disposable spreader.

nonlethal dose of bacteria over time, such as might be relevant to a chronic infection. Whatever the ultimate goal of the study, a viability assay is critical.

For experiments to be consistent, it is important always to seed with the same number of cells, culture to the same level of confluence and infect with the same dose of bacteria. Therefore, it is critical to complete these experiments to establish experimental parameters. It is possible to calculate the multiplicity of infection based on cell counts and CFU per millilitre of bacteria, and this will help to standardise your work further. By this stage, you know the CFU of bacteria to add to the cultured cells to achieve the desired effect, but you also need to know on average how many cells are present in the monolayer. To do this, seed the monolayer as normal as if preparing for the experiment, but once the monolayer appears confluent, enumerate the cells by removing the monolayer and using a haemocytometer as previously described. For small volumes, it might be best to scrape the monolayer. The multiplicity of infection is the ratio of CFU per millilitre of bacteria and cells per millilitre of the confluent monolayer. Once you have determined this, it can be used for all subsequent experiments and will allow others to repeat your work precisely if necessary.

These experiments will take time, but without them, data derived from a cell-based infection model will be meaningless, so it is important that these steps are taken before proceeding with your work. Ensuring the standardisation of techniques will make sure that any experimental effects observed in subsequent experiments are true and not the result of spurious variables caused by the failure to optimise the model properly.

5.1.4 Bacterial attachment and internalisation/invasion

Basic questions with regard to host–bacteria interactions and associated pathogenicity and virulence can be addressed by ascertaining what happens when bacteria interact will a specific, relevant cell line. Such approaches provide not only salient information about these processes but also a good start point if little is currently known. Furthermore, information gleaned from these types of experiments can be elaborated upon and used to support prior experiments (see Section 5.5).

One of the most commonly employed approaches is the gentamicin protection assay. It is a straightforward assay to conduct and can be used to answer more than the simple question of whether a particular bacterium interacts with a particular cell line. The gentamicin protection assay tells you whether the bacteria you are studying can attach to the surface of the cell line and whether it can be found intracellularly as a consequence of invasion by the bacteria or internalisation by the cell. If you are studying a pathogen, these outputs give you an idea of whether the cell line can be readily colonised by the pathogen, perhaps making it a target site for infection if the bacteria invades the cells, and therefore potentially survives intracellularly or perhaps subverts the host cell in some way. For opportunistic pathogens, an experiment might be designed to see how attachment and invasion change over time or in response to different external factors that can be added to the model (e.g. a change in pH or the presence of a particular drug). For commensal bacteria, it might simply answer the question of how readily cell are colonised and the type of response this elicits from the host.

To generate meaningful data, is it important to have identified an appropriate cell line, as described in Section 5.3. For studies with an immunological focus rather than a specific anatomical site, a monocyte or macrophage cell line could be used, such as THP-1 monocytes, which can be readily differentiated into macrophages.

The aim of the gentamicin assay is to assess bacterial interaction with cultured cells. To do this, it is important to ensure that the dose of bacteria used to infect does not kill the cells within the time frame of the experiment. You will have ascertained the lethal dose previously (Section 5.4). In terms of planning, you might want to consider whether it would be informative to use a range of nonlethal doses or to collect data at different time points. Determine this in advance of the experiment to ensure that you prepare enough cell monolayers to work with, bearing in mind the need for a negative control (not infected) and sufficient experimental replicates.

This section describes the basic setup for the gentamicin assay and offers some considerations for experimental planning. The gentamicin assay can be carried out in a 24-well microtitre plate. This volume allows for ease of sample handling. Therefore, first calculate how many cells with which to seed each well and how long to culture them to achieve appropriate confluence. During this part of the preparation, it might be good to enumerate the cells in preparation for determining the multiplicity of infection. Owing to the nature of sample processing as the experiment proceeds, it is a good idea to prepare a plate of cells for analysis of attachment, and a second identical plate of cells for analysis of invasion/internalisation. Each plate should include sufficient cell monolayers for triplicates of each sample, and to account for all time points (Fig. 5.4). Typical start points with regard to time might be 1 and 2 h infection, which is used in the example given here.

Keep the plate containing the cell monolayers in a 37°C incubator until needed. In the meantime, prepare the bacterial inoculum using the previously

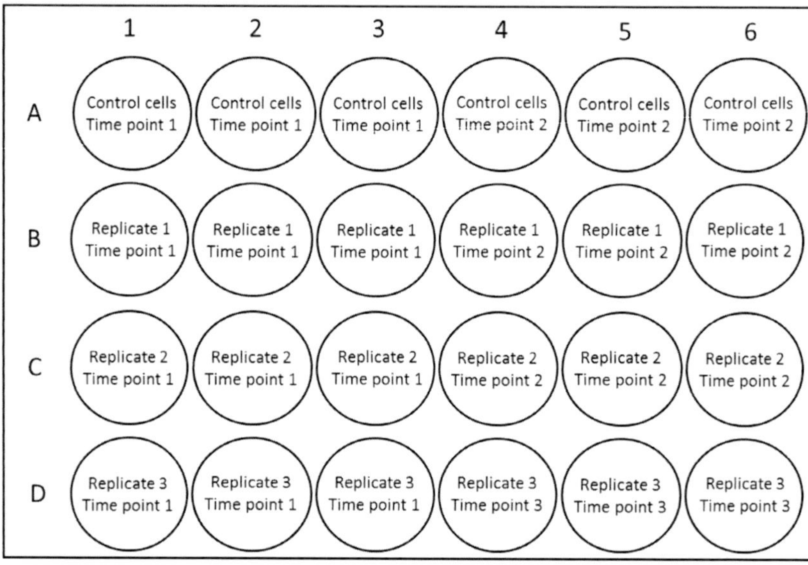

Figure 5.4 Example of layout of a 24-well microtitre plate for use in the gentamicin assay. Additional time points will require additional plates. Two identical sets of plates should be prepared: one to assess adhesion and one to assess invasion.

optimised infection doses (see Section 5.4). When preparing a bacterial dilution, use the cell culture medium as a diluent. Therefore, initially you will need to centrifuge the original bacterial culture and pour away the spent bacteriological media. Resuspend the bacterial cell pellet in the same volume of cell culture media as used to grow the bacteria. Then, prepare the appropriate infectious dose by diluting this bacterial suspension in cell culture media, making sure to work aseptically at all times. Inoculate the test well of each 24-well plate with the dose of bacteria (do not inoculate the negative control) without disturbing the cell monolayer. Place the inoculated plates into a 37°C incubator for 1 h. Remove both plates from the incubator. For the first plate (assessment of bacterial attachment), gently aspirate the medium from the wells by pipette and discard it. Wash the monolayer three times by adding sufficient PBS to cover the monolayer, discarding the PBS after each addition. Next add 1 mL of sterile distilled water to the monolayer. The osmotic shock will cause the cell monolayer to lyse. Using a 1-mL pipette tip, scrape the monolayer from the bottom of the well and resuspend it in the water. This will give you a suspension containing lysed cells and any bacteria that were either attached to the surface or internalised. Prepare a serial 10-fold dilution (10^{-1} to 10^{-6}) of the suspension using PBS and prepare a lawn plate of the sample, as described in Section 5.3. Incubate the plates overnight at 37°C; the number of bacteria recovered can be used to calculate the CFU per millilitre and the value obtained will be the total number of bacteria associated with the cell monolayer (inside and out). For the second plate, remove the cell culture medium and discard it, and wash the monolayer three times as previously described (the wash steps can be performed for both plates simultaneously). Next, add 1 mL 50 μg/mL gentamicin dissolved in sterile PBS to each well and incubate the plate at room temperature for 1 h. Afterward, remove the gentamicin and wash the wells three times with sterile PBS. Add 1 mL sterile distilled water and process the samples as described earlier. The gentamicin will have killed all of the surface-attached bacteria, and so the CFU per millilitre recovered for these samples will be the number of bacteria that were inside the cells and therefore protected from the gentamicin (check that the bacteria you are working with are susceptible to gentamicin; if not, you can use a different antibiotic). To calculate how many bacteria were attached to the cell monolayer, subtract the CFU per millilitre value for the intracellular bacteria from the CFU per millilitre for all attached/intracellular bacteria (i.e. the values obtained for Plate 2) from the values obtained from Plate 1.

The data generated from this assay will give an initial idea of how the bacteria interact with the host cells. They can be used as a basis for changing some of the parameters, such as to study different infection doses and what happens if the infection is allowed to take place for 1, 2, or 3 h. For immunological studies, the spent media from the infected cell lines can be collected and used for further analyses, such as an enzyme-linked immunosorbent assay (ELISA) (see Section 5.5). When interpreting and presenting the data, take average values for the replicate samples and calculate the standard error of the mean. The CFU per millilitre for each sample can be presented graphically to indicate differences between samples (Fig. 5.5).

By itself, the information obtained from the gentamicin assay can answer simple questions about the capacity of a bacterium to interact with a host cell and can be further developed to understand the implications of such interaction. For example, if no intracellular bacteria are recovered, it might indicate that the bacteria do not invade these particular cells or perhaps cannot survive intracellularly. If low numbers of surface-attached bacteria are recovered, it might mean that these are not the optimal cells for colonisation or that other factors are required to enhance attachment. These findings might lead you to repeat these experiments with other different cell lines from similar anatomical sites, or to change a model that used an immortalised cell line to one that uses a primary cell line.

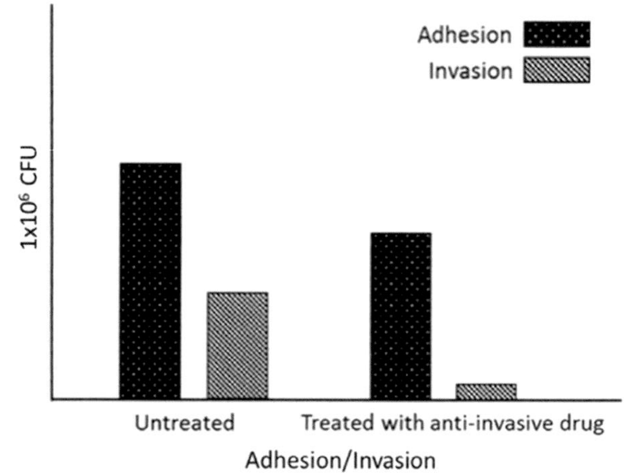

Figure 5.5 Example of how data from a gentamycin assay might be presented. In this case, the graph compares a sample that was not treated with one that was treated with a drug thought to inhibit the invasion of bacteria into the cells. *CFU*, colony forming unit.

If you have used the gentamicin assay in a series of experiments that investigated bacterial growth and antimicrobial susceptibility, for example, you can begin to build these findings into a broader understanding of host–bacteria interaction. This approach will be described in more detail in the next section.

5.1.5 Using cell culture-based infection models to support previous observations and plan the next experiments

Each experimental approach described so far in this book has limited merit when used in isolation. Many of them can be combined to give a more comprehensive understanding of the subject of interest. However, most of these experiments provide preliminary evidence for your studies that will need to be appropriately elaborated upon at a later date. Nevertheless, for an undergraduate-level project, these techniques will provide a solid foundation for robust investigation.

First, consider bacterial growth: you will have generated some data that look at the growth profile of bacteria in liquids and on solid media, perhaps with different supplements and inhibitors. With regard to pathogenicity and virulence, the growth phase can be fundamental to the expression of virulence factors and host interaction. For example, some bacteria express virulence factors primarily in the exponential phase of growth, whereas others such as toxins might be predominantly produced during the stationary phase. Therefore, knowledge regarding the time at which the different stages of growth occur for a specific bacteria could be used to assess differential virulence. A simple experiment might use the viability and gentamicin assays to investigate bacterial inocula that were collected at each different phase of growth. The findings will then relate growth to virulence and/or the capacity to colonise human cells.

Data derived from antibiotic profiling can also be used in cell culture-based models such as the viability or gentamicin assay. For example, it might be possible to determine whether a particular antibiotic inhibits attachment or invasion, or whether a novel antimicrobial can rescue infected cells from infection and subsequent cell death. The latter would rely on the study of viability for an infected cell line that is subsequently treated with a given antimicrobial agent. Parameters such as dose and time of infection could be included. Certainly, for novel antimicrobials a viability test for the cell line would have to be carried out to ensure that any therapeutic doses were not toxic to the host cell. These data can then be mapped to initial antibiotic profiling to enhance it beyond simple susceptibility profiles.

Future experimental planning can also be built on data derived from cell culture-based assays. For example, if a bacterium avidly adheres to a particular cell line, it might indicate that organism's propensity to form a biofilm. In that case, the next experiment to undertake might be a biofilm or aggregation assay (see Chapter 6). Also, consider to what cellular structures the bacteria might be attaching. For example, do the cells express fibronectin on their surface, or mucin? Might these surface-bound host proteins be ligands for bacterial attachment? With some more research into the characteristics of the cell line you have chosen to use, you can identify specific proteins that could be employed in simple protein binding assays that would allow you to suggest bacterial binding to a specific host cell protein.

From the host perspective, it might be possible to identify cell changes in response to bacterial colonisation and/or infection. In this case, by collecting spent cell culture media after infection, you could carry out an ELISA for cell-specific cytokines, or by extracting RNA from the cell monolayer you might assess expression of these cytokines (see Chapter 7). To add weight to the study, both gene expression analysis and ELISA would provide an indicator of the host cell response in terms of the switching on of gene expression and subsequent protein translation, respectively. Analysis over a given time course after infection or in response to different infectious doses would also shed light on how quickly the host cell responds and the magnitude of the response. Planned properly, these types of experiment could be undertaken in concert with a viability assay to maximise the amount of data generated from your experiments.

Of course, cell culture-based models have limitations, and even when primary cell lines are used, they are not truly representative of the in vivo response to infection. Invertebrate infection models are becoming increasingly popular as preliminary in vivo infection models to assess the virulence and efficacy of antimicrobial treatments. These will be discussed in greater detail in Chapter 10.

5.2 Notes page

Record observations and notes here.

CHAPTER SIX

Biofilm models to understand infectious diseases

Contents

6.1 Introduction

When cultured in the laboratory, bacterial isolation typically relies on agar in a Petri dish or liquid media to produce planktonic growth. However, a significant number of bacteria, both commensal and pathogenic, ordinarily grow as a biofilm. Biofilms are sessile communities of bacteria that grow attached to a given surface, known as the substratum. Biofilms can form over a several hours, eventually maturing and remaining in place for long periods of time; this can be from weeks to years, depending on the bacterium and the environment. Biofilm bacteria produce a thick polysaccharide layer that prevent the diffusion of immune molecules and antimicrobials, which makes biofilms in infection especially difficult to resolve. Chronic infections are often associated with the presence of a biofilm and have poor outcomes for affected patients. This chapter will introduce and explain how to carry out several assays to determine whether a bacterium forms a biofilm, and whether that biofilm can be disrupted.

6.1.1 Does the bacterium form a biofilm?

There is little point in using biofilm assays if you happen to be working with a bacterium that does not produce a biofilm. More often than not, it will be possible to find out this information by a literature search. However, sometimes different species of the same genus or different strains of the same species have differing capacities to form a biofilm. There are two simple assays

Bacteriology Methods for the Study of Infectious Diseases
ISBN 978-0-12-815222-5
https://doi.org/10.1016/B978-0-12-815222-5.00006-7

you can carry out to determine whether the bacteria you are studying form biofilms; these will be described subsequently.

The early stages of biofilm growth are typified by the formation of bacterial aggregates and irreversible attachment. Aggregation can be assessed over a relatively short time frame using a spectrophotometer. First, you will need to prepare a culture of bacteria in liquid media (5–10 mL) and allow it to grow for 16–24 h at an appropriate temperature, ordinarily 37°C. Next, the culture must be centrifuged to pellet the bacteria. You can use a 15-mL sterile centrifuge tube and centrifuged it at 4000 revolutions per minute (rpm) for 15–20 min. Decant and discard the spent media and resuspend the pellet in 1–2 mL sterile 3-(*N*-morpholino)propane sulfonic acid (MOPS) (30 mM, pH 7.0). Transfer 1 mL of suspension to a semi-microcuvette and read the optical density (OD) of the suspension at a wavelength of 600–650 nm, using a cuvette containing 1 mL MOPS as a blank. For this assay, the suspension must be between 0.8 and 1. If it is too high, you will need to add phosphate-buffered saline (PBS) to the suspension until it is sufficiently diluted. If it is too low, you might need to use more culture.

Once you have equilibrated the culture, place it on ice. Prepare fresh semi-microcuvettes for the assay (one for each bacterium). Vortex the cultures you have on ice and add 1 mL to each of the semi-microcuvettes (Fig. 6.1). Take an OD reading immediately ($A_{600\text{-}650}$) and then place the tubes into a rack on the bench. Take readings every hour for 3–4 h (you may need to do

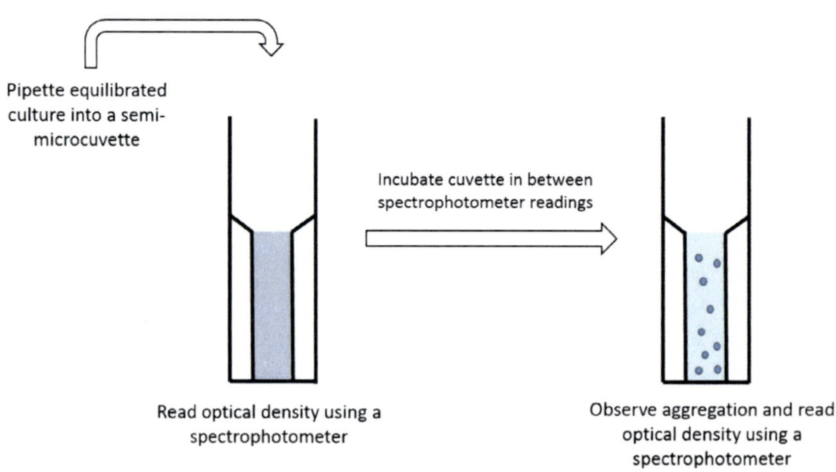

Figure 6.1 Schematic diagram of an aggregation assay. Aggregates of bacteria form in the semi-microcuvette over time and aggregation can be quantified by spectrophotometry.

this for longer). If the bacterium you are studying aggregates, you will begin to see clumps of cells forming in the semi-microcuvette and they may begin to settle at the bottom. This will result in a reduction in the OD reading over time. Once the assay is completed, it is possible to establish how much aggregation has occurred at each time point by calculating the percent reduction in OD compared with time 0. If there are obvious aggregates in suspension in the cuvette, you could also take a photograph to present alongside the calculated percentages. Bacteria that aggregate tend to also produce a biofilm.

The simplest biofilm assay is the crystal violet assay to determine biofilm biomass. This is carried out in a sterile 96-well, flat-bottomed microtitre plate in the following manner. Having previously prepared a liquid culture of the bacterium of interest, you need to wash and equilibrate it to an OD of 0.8–1 ($A_{600-650}$) as described earlier (you can resuspend it in fresh sterile media). Next, aseptically fill the wells of the microtitre with 90 μL fresh, sterile media. Fill as many wells as you need for the assay, remembering that each sample must be assayed in triplicate and that you will need a negative (un-inoculated) control, also in triplicate. Then, inoculate each test well with 10 μL of equilibrated bacterial culture. Place a lid or plate-sealer onto the microtitre plate to prevent any culture or media from evaporating and incubate the plate at 37°C overnight. If you are interested in very early attachment, you could prepare sufficient samples to test for biomass production at an earlier time point: for example, 4, 6, 8, 10 and 12 h.

After incubation, you need to stain the biofilm with a dye to determine the biomass (Fig. 6.2). If the bacterium you are studying is a prolific biofilm former, you might be able to already see a whitish film covering the bottom of the wells of the microtitre plate, but this still needs to be stained to be quantified. Tip the plate forward at a slight angle and use a pipette to remove as much of the media as possible without touching the biofilm. Use a 100- to 200-μL tip and place the tip at the point where the base of the well meets the side of the well (which is why it is a good idea to tip the plate slightly). Discard the spent media. Gently pipette 100 μL of PBS onto the biofilms and then remove it by pipette (discard the PBS). Do this three to five times. Fix the biofilm by air-drying it for 30 min. Next, add 50 μL of 0.25% (w/v) crystal violet and leave it in the wells for 10 min at room temperature. When you prepare the crystal violet solution, dissolve the powder in water and filter it through a paper filter and funnel. Gently pipette the crystal violet out of the wells and wash the wells three to five times with 100 μL PBS (using a pipette). Do this gently to avoid disturbing the biofilm. You should then see the biofilm on the bottom of the wells stained purple. Add 100 μL 7%

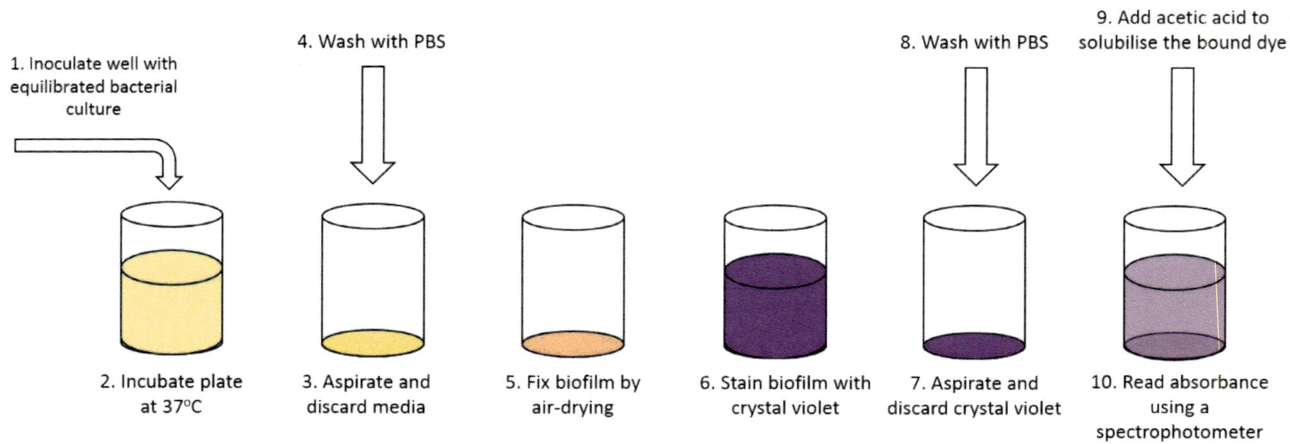

Figure 6.2 Schematic representation of stages involved in crystal violet biofilm biomass assay. *PBS*, phosphate-buffered saline.

(v/v) acetic acid to each well and gently tap the plate to mix it. You will see the liquid turn purple as the crystal violet staining the biofilm dissolves into it. Finally, read the plate in a spectrophotometer/plate-reader at A_{595}. The values you obtain indicate biofilm biomass and can be presented graphically as a bar chart. The control will give a very low background reading, which you should subtract from all of the biofilm readings.

A variation on this assay is the air–liquid interface biofilm assay. It is carried out in essentially the same way, except using a round-bottomed microtitre plate. This assay is good if the bacterium you are studying is motile and there is evidence that it typically grows at an air–liquid interface, such as the surface of a wound. When media are aspirated from the well of this type of plate, the biofilm will remain in a ring where the meniscus of the liquid was originally (Fig. 6.3). When washing, use 150 μL PBS to ensure that the biofilm is submerged and nonadherent cells are removed. When staining, use 150 μL crystal violet to ensure the biofilm is submerged, and wash with 160 μL PBS afterward. Finally, add 150 μL crystal violet to each well to dissolve the crystal violet to be read in a spectrophotometer/plate reader, as previously described. The data can be presented as per the assay using a flat-bottomed microtitre plate.

6.1.2 How many bacteria are in the biofilm and are they viable?

Determining the biomass of a biofilm using the crystal violet assay has limitations. The entire biomass is stained and includes viable and nonviable bacteria as well as other debris and the polysaccharide layer surrounding the biofilm. It gives no indication of how many bacteria are present in the biofilm or what proportion are alive. These are useful to know if your study aims to determine whether a given treatment can penetrate the biofilm and kill the bacteria in it. There are commercially available kits and dyes that enable you to determine viability, as well as traditional methods of total viable count (see Chapter 2).

The resazurin-based viability assay described in Chapter 5 can be used to assess the viability of bacteria in biofilms. A number of factors can be assessed in this way, such as that of biofilms cultured for different lengths of time or in response to different doses of a particular antimicrobial, or the same doses over a series of defined time points, or the efficacy of a new or putative antimicrobial compound. This assay lends itself to the analysis of biofilms cultured in 96-well flat-bottomed plates. When preparing the biofilms for staining, do not fix them in any way, because this can affect

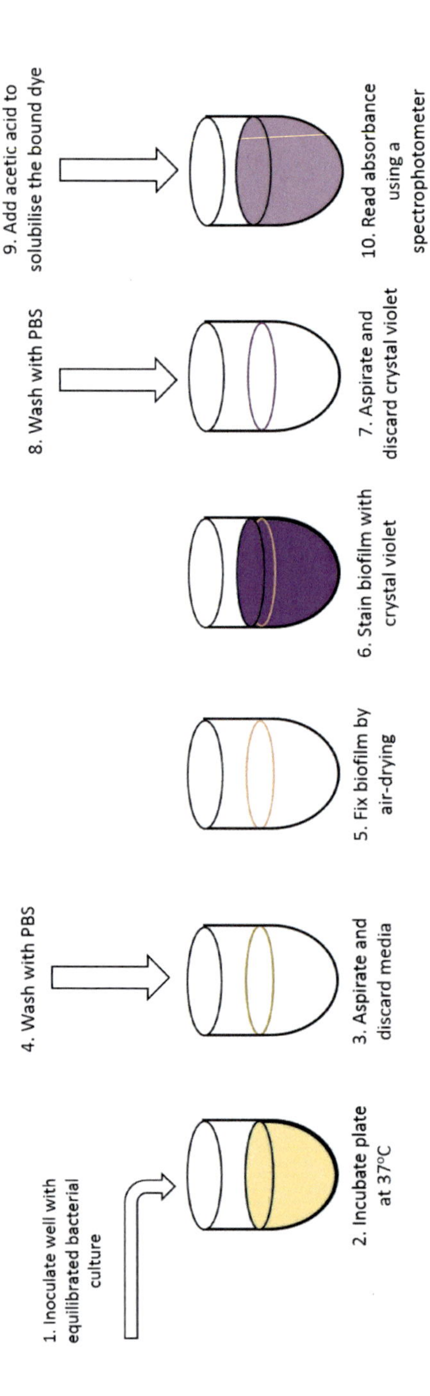

Figure 6.3 Schematic representation of stages involved in analysis of biofilm formation at air–liquid interface, staining for biomass. *PBS*, phosphate-buffered saline.

viability. Other commercially available alternatives to resazurin-based viability assay include the adenylate kinase release assay (as an indicator of membrane disruption) or adenosine triphosphate quantification as a signal for metabolically active cells using the luciferase reaction. Both of these assays can be carried out on biofilms cultured in 96-well flat-bottomed microtitre plates. Therefore, to save time, if you are preparing biofilms for a crystal violet assay, it is a good idea to prepare a replicate plate that can be used for a viability assay.

All of these assays quantify viability as a percentage of the control (for example, untreated). However, it is possible to supplement this information using microscopy to obtain an image of the biofilm, including the cells in it and which are viable. This relies on fluorescent microscopy and a combination of commercially available dyes that stain either viable or nonviable bacteria. A commonly used system is the BacLight viability stain for membrane integrity (Thermo Fisher Scientific). This uses two stains: SYTO 9, which stains viable cells only and appears green by fluorescent microscopy; and propidium iodide, which can only permeate nonviable bacteria by virtue of membrane disruption and stains DNA. The latter dye appears red by fluorescent microscopy. To use these stains, you must check that the fluorescent microscope you are using has the correct filters (see the manufacturers' guidelines). You will initially take two images, one using the green filter (excitation/emission = 494/517 nm) and one using the red filter (excitation/emission = 517–617 nm) and overlay these to produce a composite image showing both viable and nonviable cells. These images also give you a good indication of the structure of the biofilm (Fig. 6.4).

If opting for fluorescent microscopy, biofilms must be prepared differently from the 96-well plate assays. You will prepare a culture that is incubated overnight as usual and equilibrate it using OD. However, for these experiments you need to grow the biofilms on round glass coverslips placed inside the wells of a 24-well microtitre plate. First, decide how many biofilms you need to grow and then place the coverslips into the well of the microtitre plate. Add 900 μL of media to each well, ensuring the coverslip is submerged. If it floats, submerge it using a pipette tip. Next, inoculate the wells with 100 μL of equilibrated bacterial culture and incubate the plates according to the parameters you used for the 96-well assays (including all tests). Once the biofilms are ready to stain, remove the media and discard them. Wash the biofilms three to five times with sterile water and then add the two dyes (prepared according to the manufacturer's instructions). You might need to spend some time optimising the

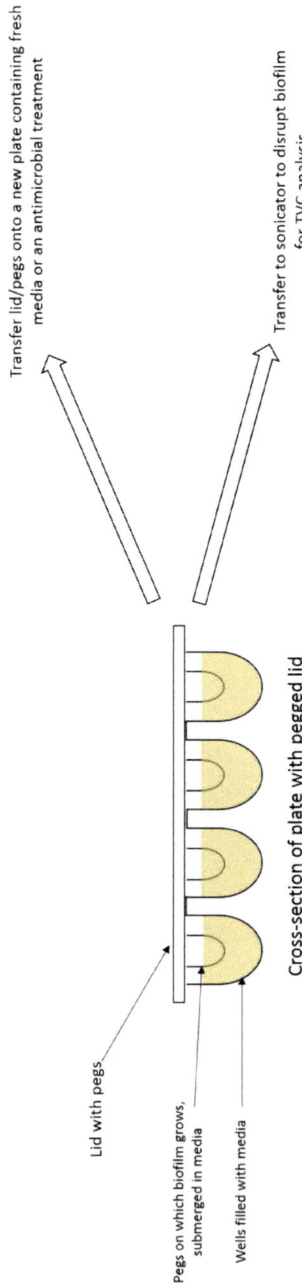

Figure 6.4 Individual and overlaid images of biofilm stained with live/dead fluorescent dyes. *TVC*, total viable count.

ratio of dyes and the length of time for staining (refer to the manufacturer's recommendations to do this). So, your first experiment might be to grow identical biofilms and test different dye ratios and staining times. Once the biofilms are stained, you will discard the stain and wash the coverslips three to five times with distilled water. Using forceps, you need to remove the coverslip from the well of the plate and transfer it to a glass microscope slide, with the biofilm facing upward. The BacLight kit comes with mounting fluid; place a very small drop of this onto the microscope slide and place the coverslip onto the mounting fluid. To hold the coverslip in place, put four dots of clear nail varnish around the edges and allow them to set. The dyes will begin to bleach when exposed to light, so once the slides are ready, put them into a container wrapped in aluminium foil to keep out the light. View the slides as soon as possible. Remember to include a scale bar on any images you produce.

An alternative way to assess how many viable bacteria are present in a biofilm is to use the total viable count (TVC) (as described in Chapters 2, 4 and 5). However, this does not provide you with any information about the structure of the biofilm or what proportion of the biofilm is not viable. To perform TVCs for biofilms, you can use a 96-well plate format, but it is advisable to use a bigger plate such as a 48- or 24-well plate for ease, adjusting volumes of media and inoculum accordingly. Culture biofilms as per the 96-well plate assay and discard the spent media. Wash the plate three to five times with PBS as per the 96-well assay, but after the final wash take a sterile pipette tip and scrape it over the entire surface of the well to remove all of the biofilm (you most likely will not be able to see the biofilm). Then, add 0.1–1 mL media (depending on the size of the well you used) and prepare a serial dilution to plate out and enumerate them (see Chapters 2, 4 and 5). If you wanted to assess biofilm structure in conjunction with this approach, you could culture biofilms on coverslips (as for BacLight staining) but stain them with crystal violet for 5 min and wash as per the 96-well assay protocol before mounting them onto a glass slide and observing them by light microscopy.

6.1.3 Applications of simple static biofilm models

The biofilm models described earlier can be used to answer a variety of research questions. For example, it would be possible to use the crystal violet assay to determine how quickly a biofilm might establish. One limitation of the crystal violet biofilm assay is that it relies on batch culture and therefore allows analysis only of early biofilm, This feature can be exploited for studies investigating early biofilm growth. This type of experiment could establish biofilm development at several different time points over 72–96 h (to give time for a final analysis of biomass and TVC). It might be expected that the overall biomass would increase and the structure of the biofilm move from discrete aggregates to greater coverage of the substratum (which could be assessed by microscopy). The media would become increasingly deficient in nutrient over time, with a likely buildup of waste metabolites that could be toxic to the biofilm. In this case, it would perhaps be good to perform a second experiment simultaneously, in which the spent media are removed every 24 h and replaced with fresh media to assess how this affects biofilm development.

Alternatively, the crystal violet biofilm assay can be used to assess whether known or uncharacterised compounds with suspected activity are effective in preventing biofilm formation and/or disrupting an established early biofilm. The principle for this type of experiment is similar to that of the microbroth method to determine minimum inhibitory concentrations (MIC). As such, a plate identical to that for the MIC experiment could be prepared and the crystal violet assay used to stain any biofilm in the wells of the plate. This would determine the minimum biofilm exclusion concentration for the antimicrobial compound and is the lowest concentration of an antimicrobial that prevents biofilm formation. Usually this is higher than the MIC value. To assess the ability of an antimicrobial to disrupt early established biofilm, biofilms can be cultured for 24, 48 and 72 h in a 96-well plate and then treated with the antimicrobial agent(s) (for different specified lengths of time: one treatment time per biofilm). After treatment, the remaining biomass would be determined by staining with crystal violet and the results compared with an untreated control to establish any reduction in biofilm biomass as a consequence of treatment. As with previous examples, these experiments could include simple light microscopy and TVC to add to the biomass data. Straightforward statistical analysis using analysis of variance and post hoc comparison (e.g. Bonferroni or Tukey) would help you to determine whether any reduction in biomass was attributable to the antimicrobial treatment and whether any particular length of time was critical for this reduction to occur.

6.1.4 Other types of static biofilm model

The crystal violet model is quick and easy to use; it generates useful data as a good starting point for studying bacterial biofilms. However, it has limitations: biofilms are cultured in batch mode and give only end point

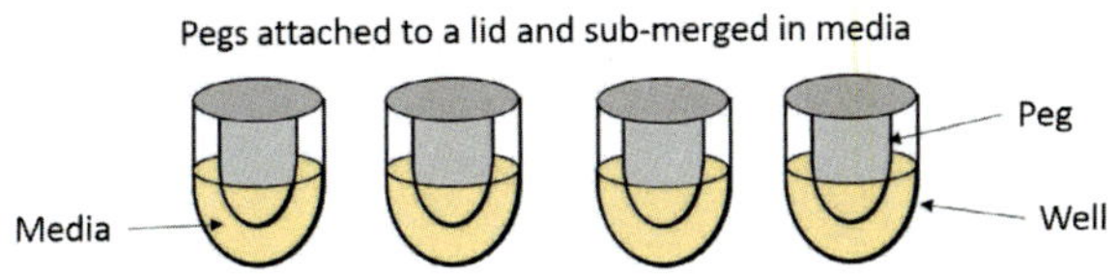

Figure 6.5 Schematic representation of the Calgary Biofilm Device.

measurements; and the crystal violet stains other debris and to some degree the plastics in which the biofilm is cultured, which can cause problems with reproducibility. This section introduces three other static biofilm models and explains their advantages and disadvantages. If these are available to you, they are worth investigating. However, some are expensive, and so the crystal violet assay may remain the best option.

The Calgary Biofilm Device is similar to the crystal violet assay in that it uses a 96-well format to assess biofilm growth, giving end point measurements and being versatile enough to allow for testing of antimicrobial agents (Fig. 6.5). Specialised 96-well plates are used that have lids to which 96 pegs are attached. When put on the microtitre plate, these sit within the wells. The wells of the plate are filled with media, much like for the crystal violet assay, and biofilm grows on the pegs. You prepare this assay in the same way as a crystal violet assay but the end point analysis is different. Instead of staining the biofilm, the pegs are removed and sonicated to dislodge the biomass, which is then enumerated by TVC. Pegs can be removed at different time points throughout the experiment, and although biofilms are grown in batch culture, fresh or different media (perhaps containing antimicrobials) can be introduced by transferring the lid and pegs to a fresh microtitre plate containing different media and/or supplements or antimicrobials.

The biofilm ring assay is another microtitre-based assay that uses paramagnetic beads to assess early-stage biofilm formation (Fig. 6.6). These assays also require a specialised magnetic device and scanner, which are bought along with the system. This assay does not require staining as biofilms are scanned. It is a useful assay to assess the efficacy of compounds that might have anti-biofilm activity, which is those that prevent bacteria from aggregating and adhering to a substratum. Anti-biofilm agents are being explored as an alternative to traditional antimicrobials. Often, they do not kill bacteria but instead impair biofilm formation. Briefly, the wells of a microtitre plate are filled with media and are inoculated in the same way as for the crystal violet assay. To each well is then added an aliquot of paramagnetic beads. The principle of the test is that if bacteria form a biofilm,

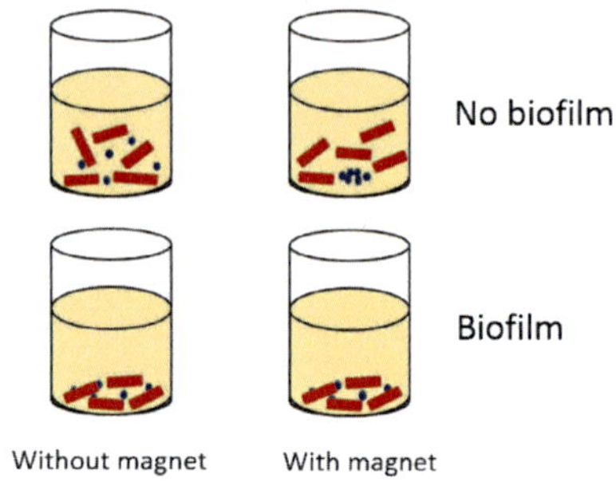

Figure 6.6 Schematic representation of biofilm ring model.

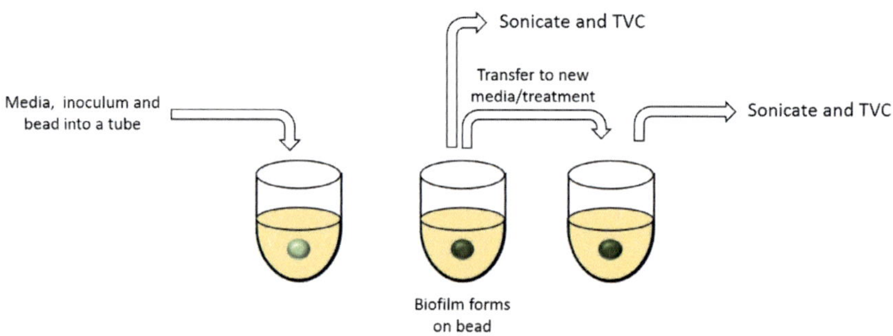

Figure 6.7 Schematic representation of biofilm bead assay. *TVC*, total viable count.

the magnetic beads cannot form a ring on the bottom of the plate when exposed to the magnetic device. If there is no biofilm, a ring of paramagnetic beads is observed. Therefore, this assay provides rapid end point information about biofilm formation.

The biofilm bead method is a newer assay and does not use a 96-well plate nor does it rely on staining biofilm biomass. The general principle will be described here, but it is recommended that you refer to the original publication for fuller experimental details (Konrat et al., 2016). This assay is also relatively inexpensive and requires little specialist equipment (Fig. 6.7). Briefly, biofilms are cultured, adhered to glass beads and enumerated by TVC. Glass beads (3–5 mm) have been used; these must be thorough washed using detergent and water before being sterilised by autoclave (Konrat et al., 2016) and are used only once. The wells of a 24-well microtitre plate are filled with 1 mL media and inoculated with a culture that was equilibrated overnight (as described previously in this chapter). Beads are then added to each well. The plate is placed into a moisture chamber (a plastic box with a clip lid will do the job) before being incubated at 37°C for 24 h with rotation at 150 rpm in a shaking-incubator. The following day, the beads are carefully removed from the wells using sterile forceps and transferred into 2 mL microcentrifuge tubes containing PBS. The samples are sonicated to dislodge the biofilm from the beads and bacteria enumerated by TVC. Like the Calgary device, this assay allows for biofilms to be transferred into fresh media or into media containing an antimicrobial for efficacy testing. It also allows for biofilm growth to be assessed over time in much the same way as the crystal violet assay. An advantage of this model is that biofilms cultured on beads can also be assessed by confocal microscopy or scanning electron microscopy to observe the biofilm structure directly (see Konrat et al., 2016 for the detailed protocol).

A different type of biofilm assay that uses no type of microtitre plate system is the Lubbock method. This model might not be suitable for all biofilm analyses, but it is good if you are interested in culturing biofilms over longer times at an air–liquid interface. As such, it lends itself well to the study of chronic infection in which biofilms are present for many months or years. The Lubbock method also allows media to be changed for long-term growth experiments, and the position of the biofilm enables topical treatments such and antimicrobial dressings to be applied. Biofilms are fed using nutrients that are immobilised in agar in a Petri dish (Fig. 6.8). Thus, essentially, you could use a nutrient agar plate as a starting point and perhaps choose different media or nutrients for later experiments to investigate how this might affect biofilm growth. A 25-mm filter membrane (0.2-μm pores to prevent bacteria from passing through) is placed on top of the agar and inoculated with 20 μL of an equilibrated bacterial suspension. The lid of the Petri dish is put into place and the biofilm is incubated at 37°C. The biofilm will grow on top of the filter disk.

The filter disk can be removed from the plate and placed into 1–5 mL sterile PBS (sufficient to cover the filter paper disk) in a sterile disposable plastic tube and vortexed vigorously to remove the biofilm. Bacteria are then enumerated by TVC. This method is best-suited to study established biofilm rather than early biofilm formation, and so it is advisable to allow the biofilm to grow for a minimum of 48 h in the first instance. Fresh media can be introduced to the system by moving the filter on which the biofilm is growing onto a fresh agar plate. This stops nutrients from becoming depleted over a long time. It is possible in this manner to change the media or nutrients on which the biofilm is growing, if this is an aspect of biofilm growth in which you are interested. Topical treatments can be investigated by applying, for example, antimicrobial wound dressings or antibiotic disks to the surface of the developed biofilm for a set period of time before

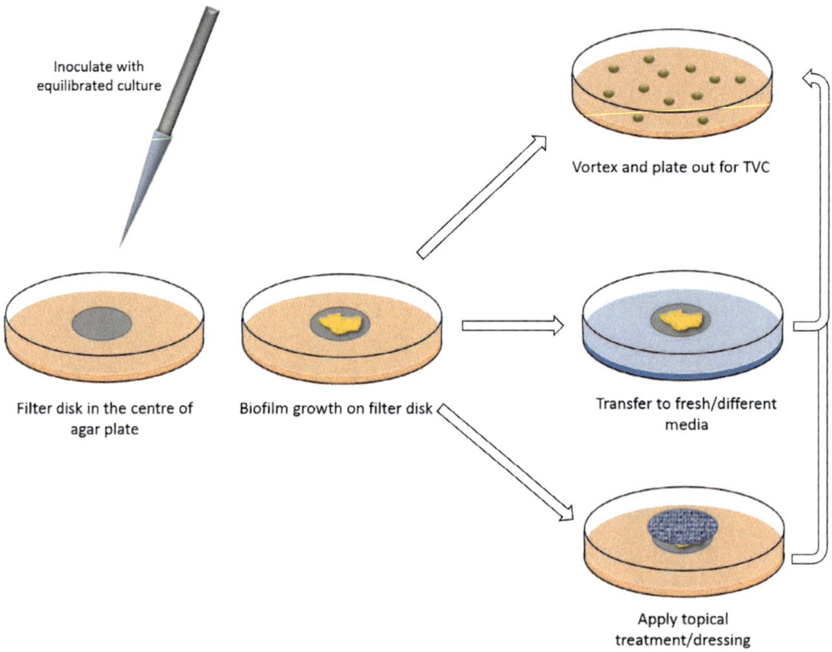

Figure 6.8 Schematic representation of Lubbock biofilm model. *TVC*, total viable count.

enumerating by TVC to assess the effect of treatment on the number of bacteria in the biofilm.

6.1.5 Studying biofilms under conditions of flow

A number of models exist that allow biofilms to be studied under conditions of flow. This is important because most environments in which bacterial biofilms are found are not static, even within the human body. This section describes some of the best-known flow devices for studying biofilms. Many are available commercially, but the costs can be prohibitive; others can be constructed in part by the user. Consider whether your studies would benefit from using a flow device, and whether you have any of these items at your disposal during your research project.

Biofilm flow devices work in a variety of ways that might make them more or less relevant to your research. For example, the Robbins device and modified Robbins device consist of an oblong metal tube with a number of holes along one face edge into which there is a series of threaded holes into which pegs are screwed (Fig. 6.9). Coupons are attached to the ends of the pegs; and one in place is in contact with media that flow through the

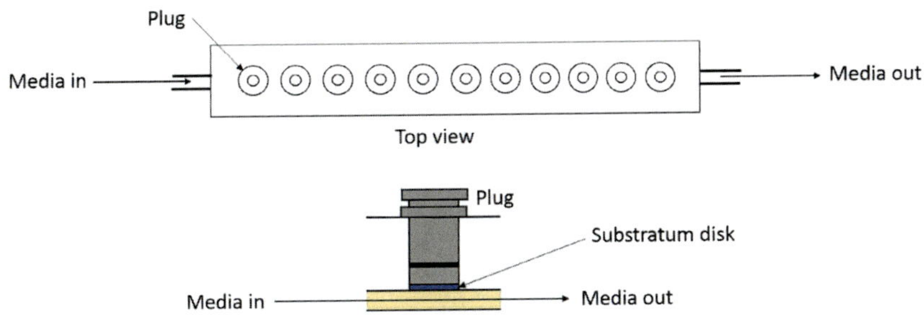

Figure 6.9 Schematic representation of Robbins biofilm device.

tube part of the device. The device is attached to a peristaltic pump to drive the flow of nutrients. Each peg can be removed without disrupting the flow of media or the ongoing experiment, and several biofilms can be cultured under the same conditions simultaneously. Coupons can be collected from the pegs and biofilm enumerated by TVC, or analysed by microscopy. The device can be set up and an experiment conducted for long periods of time providing there are sufficient media to flow through the system (more fresh media can be added as the experiment progresses). Therefore, this is a useful device for studying established biofilms. With regard to specific projects, it is important to consider what factors would make this a useful tool. Biofilms are submerged by the media at all times, and the flow can be varied depending on the environment you are trying to mimic. It is possible to look at a time course for biofilm development, but it is not easy to include treatments that cannot be incorporated into the media that are feeding the biofilms. Thus, two separate experiments would need to be performed to test whether a disinfectant impaired biofilm formation, for example.

Flow systems are also available that allow bacteria to be grown at an air–liquid interface rather than being submerged. These might better represent oral or wound biofilms, compared with those of the gastrointestinal tract, for example. One widely used system is the constant depth film fermenter (CDFF). This allows multiple biofilms to be grown on coupons that are located on a central rotating disk. Biofilms are fed by flowing media into the top of the device so that it drips onto the growing biofilms; a fixed scraper blade scrapes the top from the growing biofilms as they pass beneath it, ensuring a constant depth of biofilm (Fig. 6.10). It is possible to remove the coupons from the device as the experiment progresses, to analyse the biofilm composition and structure. Many studies have produced excellent confocal images of complex bacterial biofilms cultured this way. Like the

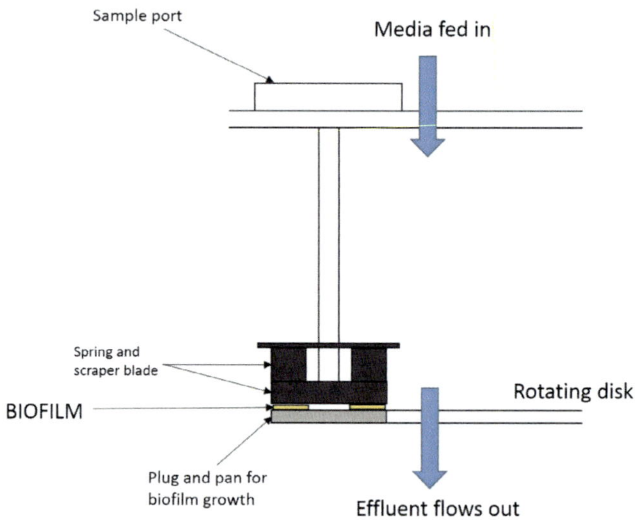

Figure 6.10 Schematic representation of constant depth film fermenter.

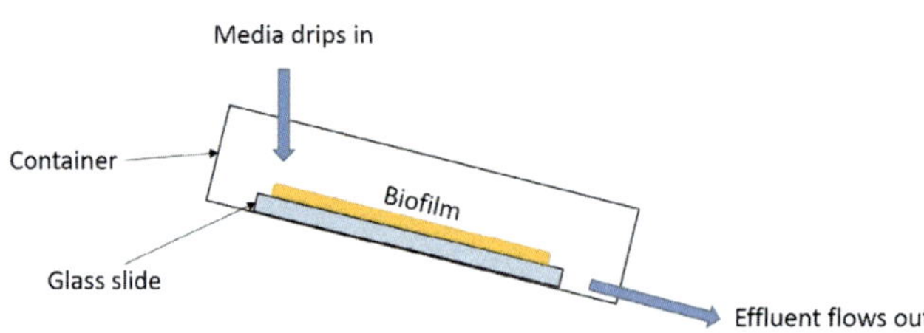

Figure 6.11 Schematic representation of a drip-flow reactor.

Robbins device, the CDFF is useful for studying established biofilms akin to those seen in chronic infection; similarly, it is not really designed to test antimicrobials.

Drip-flow reactors can also be used to culture biofilms at an air–liquid interface. Drip-flow reactors are available commercially and can be purchased or they can be constructed in-house by the user. The following paper explains how to use a drip-flow reactor (Goeres et al., 2009). The principle behind this device is that biofilms are cultured on a substratum held at an angle over which fresh media flow by dripping through an inlet portal by means of a peristaltic pump (Fig. 6.11). Waste media flow out of the device by gravity, owing to the angle at which the biofilm is grown. Devices can be used to culture single or multiple biofilms, so that it is possible to prepare replicates. Where separate reactors are present in one device, it is possible to remove biofilms at given time points. The rate of

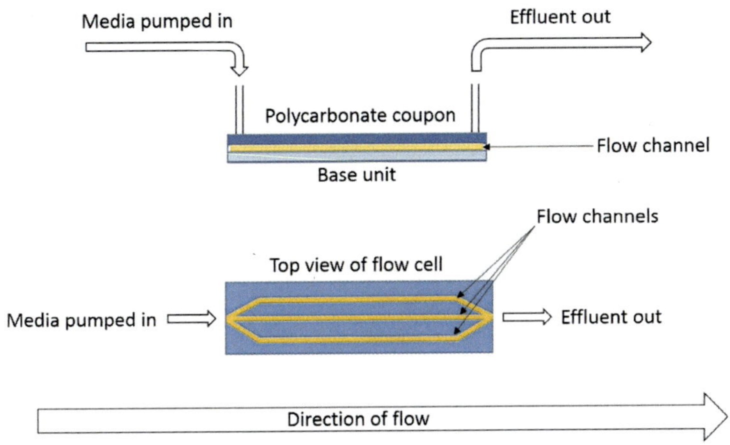

Figure 6.12 Schematic representation of flow cell.

flow can be adjusted to best represent the environment you are trying to mimic. Biofilms cultured in drip-flow reactors are commonly enumerated by TVC, but the substratum (e.g. a glass slide) can be removed and stained for microscopy. Like previous models, these devices are good for studying established biofilms, and once cultured, biofilms can be removed and treated with antimicrobial agents: for example, by immersion into a solution or by applying a topical treatment. This will ascertain how tolerant an established biofilm is to any given antimicrobial intervention.

Smaller flow cells are available that are designed for real-time visualisation of biofilm development by microscopy. These types of flow cells are available commercially or they can be constructed by the user; however, this requires some degree of technical expertise (see Tolker-Nielsen and Sternberg, 2011 for how to construct a flow cell). Like other systems, media flow through the device by means of a peristaltic pump (Fig. 6.12). Each flow cell has three channels, which means that three identical biofilms can be simultaneously cultured, each submerged in the flowing media. Again, consider the biofilm model you need relevant to the organism and system you are trying to mimic. If you gather information about biofilm development using a flow cell, the next experiment might be to flow through a compound with disinfectant properties to see how effective it is at removing established biofilm, and how quickly. This could be observed and imaged in real time.

A flow device specifically aimed at modelling wound biofilms emerged as a tool for the study of biofilm development under conditions of flow (Fig. 6.13).

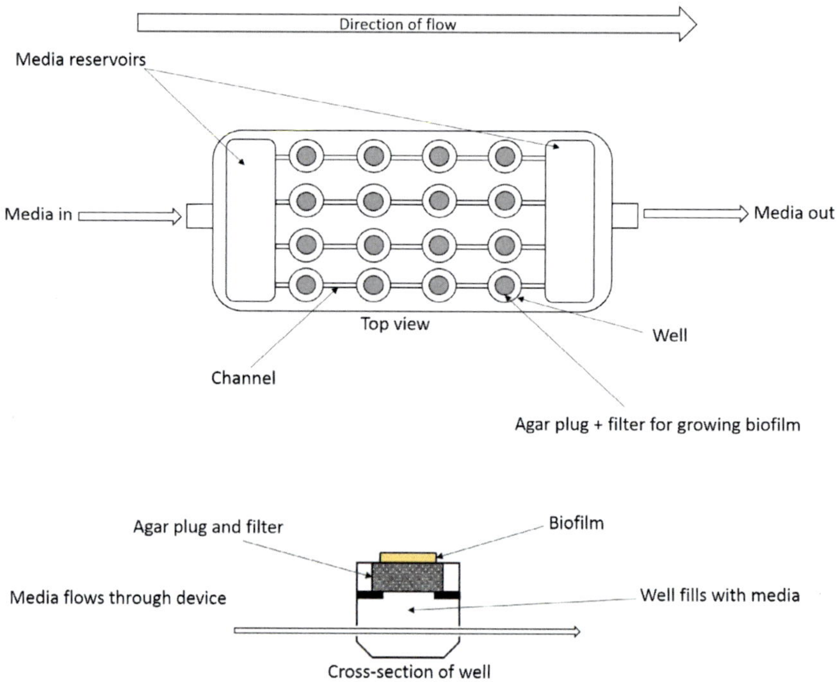

Figure 6.13 Schematic representation of the Duckworth Biofilm Device.

The Duckworth Biofilm Device (DBD) allows for the growth of 12 biofilms over three separate channels. Similar to the Lubbock method, biofilms are cultured on top of 0.22-μm filters (10 mm in diameter) on top of plugs of noble agar that are fed nutrients from channels that flow beneath it. Like other flow devices, media are pumped through the system using a peristaltic pump via an inlet and outlet port. The agar plugs are permeable to nutrients, and so the biofilm is fed in this manner.

The DBD aims to represent the wound biofilm environment better, because wound biofilms are similarly fed from beneath by wound fluid and grow at the surface of the wound exposed to the air, much like the filters in the DBD. Whereas the device can be fed by only one type of media at any one time, the three separate channels allow for the growth of three different bacteria without risking cross-contamination. Filters can be removed from the device as the experiment progresses without disrupting the flow of the experiment, and it is possible to apply a topical antimicrobial such as a wound dressing to the biofilm. The device can be used to study early stages of biofilm growth and established mature biofilms. Bacteria are recovered from filters by vortexing and TVC, but glass coverslips can be used in place of filters to analyse biofilm structure by microscopy. This device is not commercially available but can be printed three-dimensionally. Details of this and how to set up and run the system can be found in the following publication: Duckworth et al. (2018).

When it comes to deciding whether a biofilm flow device is useful for your research, you need to think carefully about whether flow is relevant to your study and what sort of questions you are looking to answer by employing a biofilm flow device. Given the availability, cost and technical expertise required to use some of the available devices, you must also consider whether the equipment is available, whether you will be able to use it effectively and whether you can afford it according to your project budget. There might be equipment available in your place of study, or your project supervisor might have contacts who have access to the equipment you need.

6.2 Notes page

Record observations and notes here.

CHAPTER SEVEN

Gene expression analysis

Contents

7.1 Introduction

The ability to investigate gene expression within bacterial isolates as a way to detect the presence or absence of genes or to target specific genes in response to experimental conditions has become an integral part of microbiological research. This approach allows a high degree of sensitivity and can provide insight into the regulation of bacteria under varying conditions. The equipment and techniques to measure gene expression are widely available and can be adapted to answer many experimental questions. Although it is limited in the sense that changes in gene expression do not always translate into a phenotypic change, this technique provides the researcher with a great deal of useful information. This chapter offers details on the basics of sample preparation and extraction, methodologies to allow the successful measurement of gene expression and information on how best to analyse the data generated by these techniques.

7.2 Sample preparation and DNA extraction

Before beginning the physical preparation of bacterial samples for DNA extraction, it is important to consider the manner in which these extractions should be carried out. Ideally, the extractions should be performed in an organised and clean fashion, because the contamination of samples could lead to erroneous results, wasting both time and reagents. It is good practice before starting extraction to have a clear bench that

Bacteriology Methods for the Study of Infectious Diseases
ISBN 978-0-12-815222-5
https://doi.org/10.1016/B978-0-12-815222-5.00007-9

has been cleaned (a standard disinfectant such as 70% ethanol can be used for this) and to check that all reagents and equipment that you will need are available on the bench close to where you are working. It is useful to make sure that you have the correct volume pipettes ready. These can also be wiped down before starting an extraction. Check whether the laboratory has an ice machine before starting polymerase chain reaction (PCR), because many PCR reagents need to be kept cool while in use. Set up a working ice bucket on your bench in a place that is easy to reach. Another essential piece of equipment when carrying out PCR is gloves; these can prevent contaminating samples and protect you from the chemicals being used, so ensure the correct size gloves are available on the bench and change into fresh ones whenever it is possible that the gloves might have been contaminated. A typical setup of a bench before PCR is carried out is shown in Fig. 7.1.

As a first step in the PCR process, once all reagents have been ordered and the bench area has been prepared, a DNA template will need to be extracted. There are several different ways in which it is possible to obtain DNA from bacterial samples, including by bead beating or vortexing in addition to using a lysis buffer. For these methodologies and more detail on

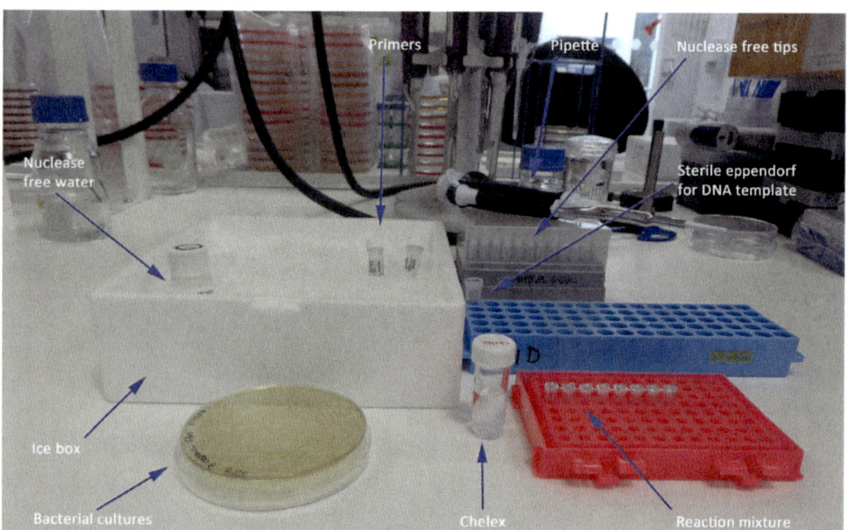

Figure 7.1 A bench setup ready for the DNA extraction process. Note the ice bucket for reagents and waste pot for used tips and tubes. It is important to have everything in a place and easy to reach to provide a logical workflow and minimise the risk for contamination.

PCR, see Maddocks and Jenkins (2017). In this section, a rapid, cheap and easy method is outlined that involves extraction using Chelex resin.

The basic Chelex method is to:

1. select several colonies of bacterial culture from overnight growth on an agar plate using a sterile inoculating loop (enough that you can see them on the end of the inoculating loop);
2. resuspend the bacterial colonies in 200 μL Chelex in a sterile container such as an Eppendorf;
3. heat the mixture to 95°C for 15 min. This should be long enough for bacterial DNA to be released from the cells;
4. after incubation, spin down the mixture at 13,000 revolutions per minute (rpm) for 3–5 min; and
5. carefully remove the supernatant from the top of the mixture and place it into a sterile Eppendorf (this supernatant contains the template DNA).

When handling Chelex (for example, when transferring it in aliquots to sterile containers), it is important to keep agitating it. Chelex is dense and if it is not mixed immediately before pipetting, the Chelex will sink to the bottom. It is easy to see this happening if you keep it in a clear-walled container, so check before pipetting. When heating the bacteria, a heating block or waterbath can be used, and if a shake option is available on the heating block, this can be turned on for 15 min incubation. Lots of methods are available for extracting DNA; this is just one example of a straightforward method that can be used as a starting point.

Once the extraction is complete, the template DNA that has been extracted can be stored until it is needed to run a PCR. If storing for a short time, the DNA can be kept at 4°C. If storing it for a prolonged amount of time, it is best to store it at −20°C.

If carrying out DNA extraction and PCR for the first time, it can be useful to have a hard copy of the protocol within sight while carrying out the procedure. Each step can then be checked off as the process progresses, which reduces the likelihood of a step being missed or accidently repeated.

Thus, following the DNA extraction, the template DNA needed for PCR is ready. However, in addition to needing DNA to carry out a PCR successfully, you will need to make sure several other components are included in the reaction mixture. If any of these components is missing, the PCR is unlikely to work even if the extracted DNA is present. The list for the reaction mixture given here provides an outline of all necessary components that can then be tailored for specific experiments:

1. Template DNA: This can be extracted in house (see previous discussion) or purchased from a manufacturer.

2. Primers: These can be ordered and kept frozen. A working stock of 100 pmol/mL should be aliquoted out and used for experiments to prevent contamination of the original stock. How to design and order primers is covered in more detail in Section 7.3.
3. $MgCl_2/MgSO_4$: These act as enzyme cofactors during the PCR, improving specificity of annealing if used in the correct concentrations. Buffers may be purchased that already contain these; they are often used and are available commercially. If making a buffer in house, a range of concentrations can be used during the initial PCRs to optimise the process.
4. deoxyribonucleotide triphosphates (dNTPs): This mix of the four bases is normally available commercially but it is best to check their concentration when purchased and aim for a stock of 100 mM (upon addition into the reaction mixture, these are then diluted).
5. Reaction buffer: Usually available as a solution containing the enzyme at a concentrated level, which is diluted upon addition to the reaction mixture. Ensure you read the instructions when purchasing it so that it is correctly diluted and check the contents (e.g. the Mg^{2+} content), as mention earlier.
6. Polymerase: For the reaction to work, a polymerase will be needed. Various types are available that can be used depending on whether you are amplifying long or short sequences; a Taq polymerase is a good starting point for most reactions.
7. Nuclease-free water: This can be purchased from various suppliers and sometime is included in kits or with the enzyme (Fig. 7.2). It is an essential part of the reactions because if any nucleases get into your reaction mixture, they could degrade the genetic material, leaving you with a failed PCR. As with any of these components, it is best to aliquot it if it is going to be used on multiple occasions, to minimise the chance of contamination.

Often, when making up the reaction mixture it will be for several reactions, in which case it is worth making up slightly more mixture than is needed on paper, because some of the mixture will be lost through pipetting. When it comes to adding the primers or nuclease-free water to the reaction mixture for the PCR, it is likely that very small volumes will be used (1–5 μL), so ensure that you can effectively manipulate these small volumes, because if any one of the components is missing from the mixture, the reaction might not work. When using small volumes, it can be helpful to pipette the liquid onto the back of the container in which the mixture is being made. This allows a visual check of the liquid that has gone in. If using this method,

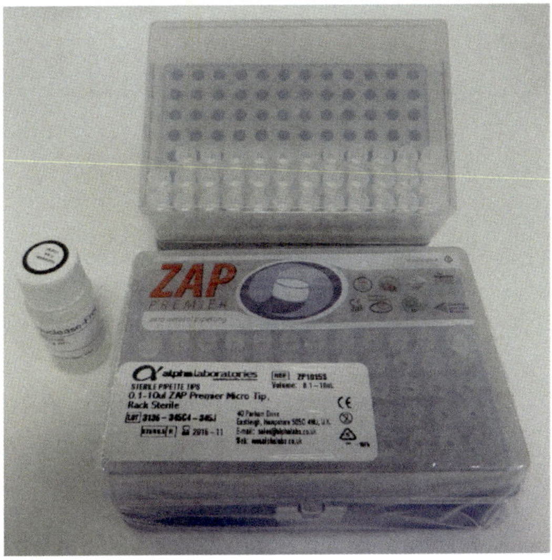

Figure 7.2 Nuclease-free water and pipette tips that can be used to protect solutions that are being handled from contamination.

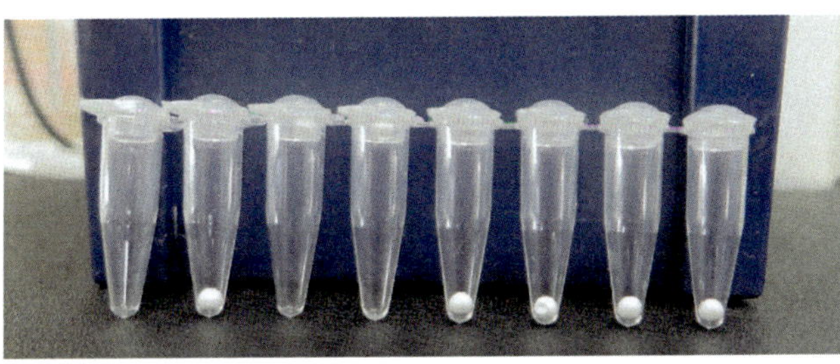

Figure 7.3 Polymerase chain reaction (PCR) beads containing the necessary components needed for a successful PCR. Primers, nuclease-free water and template DNA would need to be added to these before the reaction could take place. In several tubes, the bead has stuck to the top. Make sure the bead is at the bottom of the tube before opening the lid.

make sure the sample has all moved to the bottom of the tube before the reaction mixture goes into the PCR machine. This can be achieved by capping the tube or plate and gently centrifuging for a few seconds.

Reaction mixtures can be bought premade from various suppliers such as Fisher Scientific (fishersci.co.uk) (Fig. 7.3). When using a premade mixture, often only the primers, template DNA and nuclease-free water need to be added. When using a premade reaction mixture, check the manufacturer's guidelines.

7.3 Primers

Before it is possible to use the DNA that has been extracted, ensure that you have appropriate primers that amplify the gene of interest. Primers are DNA sequences often designed to complement an area of DNA within a gene of interest. Two primers will bind to two individual regions, and the area between these two primers will be amplified during PCR, providing a PCR product that can then be visualised and analysed (see Section 7.4).

It is possible to design your own primers. In the case of well-researched genes, sometimes it is possible to find primer sequences for the gene of interest in the bacteria being used in a published article. This means these sequences can be used instead of designing your own. In either case, it is important to make sure that primers you design or use are well-designed. It is worth taking time at the design stage to avoid problems during PCR, such nonspecific priming, secondary structure formation or non-priming.

To design primers, there needs to be an area within a gene or genes of interest that can be amplified. The template DNA you use to design a primer for those genes depends on which bacteria are being investigated. If the bacteria being researched are fully sequenced, this can make the job of designing primers easy, because it will be possible to check that the sequence chosen for the primers is within the genetic code of the bacteria. If the bacteria being used are not fully sequenced, it is possible to use online databases to give partial sequences for what should be in the bacteria of interest. There are also likely to be fully sequenced bacteria of the same species as the those being used in freely accessible online databases, but it depends on which bacteria are being researched. These online sequences can be used as a template for the primer design, but in both cases where the sequence for the bacteria being investigated is unavailable, the risk is that there may be absent or changed sequences that render primers designed on the basis of another bacterium of the same species ineffective.

When designing the primers, it is important to understand what is going to be amplified. It is a short section of sequence within a particular gene, a whole gene or several genes within an operon. For very long sequences (greater than 10 kb), specific enzymes might be required for the reaction, because it can be more difficult to amplify these sequences. For the shorter sequences, it should be straightforward. When amplifying within a gene, try to make the product larger than 100 base pairs (bp). Then, looking at the available gene sequence for the gene of interest, pick a starting point for the forward and reverse primers. Ideally, the section selected should have a guanine-cytosine (GC) ratio of around 50%. A GC content higher than this can lead to a high annealing temperature within the PCR. When selecting the region to be amplified, take care not to select an area with long runs of one base or highly repetitive areas (e.g. GCTTTTTTTTTTTTTTGTTA) because these are prone to forming secondary structures.

When designing the primers for a PCR, everyone has a particular method for finding the DNA sequence and generating primers. As a starting point, online databases can provide bacterial gene sequences (which act as the template for primer design) such as the National Centre for Biotechnology Information (NCBI) database (https://www.ncbi.nlm.nih.gov). There are also species-specific databases for some bacteria such as the *Pseudomonas* database (www.pseudomonas.com), where you can search directly for gene sequences of interest from *Pseudomonas* species.

For using the NCBI website to design primers, the following is a brief guide on how to navigate the site. Once at the NCBI homepage, select Nucleotide from the drop-down menu on the top left-hand side of the screen (Fig. 7.4).

This will take you to the nucleotide database page (Fig. 7.4), which will give you access to sequences from several sources and provide the maximum likelihood of finding the sequence of interest. It is possible to search these databases using the gene name and organism name (as show in Fig. 7.5). It is also possible to use the filters on the right side of the page to narrow the search to a specific set of parameters such as organism or species.

Once the search is complete, it will generate a list of genome sequences, some of which will not be the one of interest but that may contain the information you need. If using the bacterial name and gene name as in the example of Fig. 7.4, the list generated will often contain strains and species that are irrelevant to the search. The list returned should look something like the one provided in Fig. 7.6.

Look through the list that has been generated until you find the correct link. Follow the link and it should generate a page that contains all of the basic genome information about the bacterial strain selected (Fig. 7.7). Underneath this should be a list of the specific genes within the genome provided.

A lot of information is provided within these pages. Scrolling to find the gene of interest could take a lot of time. Probably the easiest way to see whether the gene is there is to use the Find function within the computer's browser. This should direct you to the gene of interest, where there are several links (Fig. 7.8).

Select the gene link; following this link will generate a new page that shows the gene of interest and can be used as the template for primer design (Fig. 7.9).

Figure 7.4 National Centre for Biotechnology Information homepage with drop-down menu showing options including link to nucleotide that will allow you to search for genome sequences within GenBank.

Generally, the nucleotide code is pasted into a template box on these primer design webpages. The nucleotide code should be entered unbroken and with no annotation into these boxes or it will not work. After the code has been entered, the parameters need to be set. These tend to include things such as the maximum and minimum size of the product to be generated, the annealing temperatures and the GC content. To increase the chances of a successful PCR and efficient amplification of the target, some rules of thumb should be followed when setting the parameters in primer design:

1. The first parameter to consider is the length of the primers themselves. Generally, the ideal length is 18–25 nucleotides. This should provide specificity and efficient amplification.

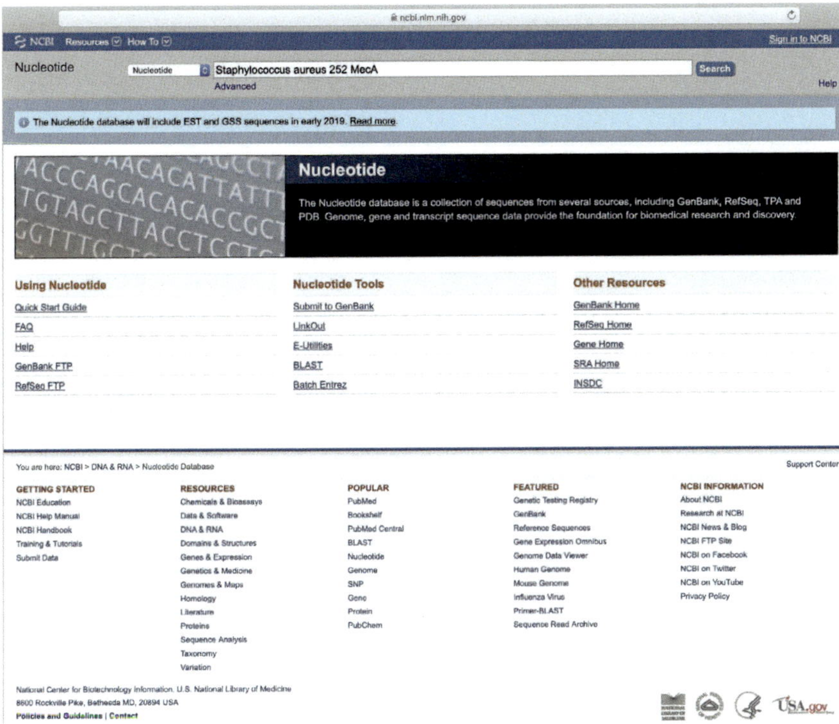

Figure 7.5 The nucleotide database page, which provides access to numerous sequences from a variety of sources. The search for the organism and gene of interest can start here. Enter the organism and gene into the search bar and press Search.

2. The GC content should be 50–55%.
3. Within the primer sequences, try to avoid repeats of bases of one type (try for no more than three of any one base in a row) to prevent secondary structures from occurring. Try to avoid more than three bps of intra-primer homology to prevent double-strand structure formation. Finally, avoid homologous regions between primers in a pair to prevent them from binding to one another (primer dimers).
4. The melting temperature of the primers should be as similar as possible, preferably within 2°C of one another. There are two primers in each reaction, and they will need to work under the same reaction conditions. Primer design websites and manufacturers will predict the primer melting temperature, but it also possible to estimate the melting temperature using the calculation:

$$Tm\,(^\circ C) = 2\,(A + T) + 4\,(G + C)$$

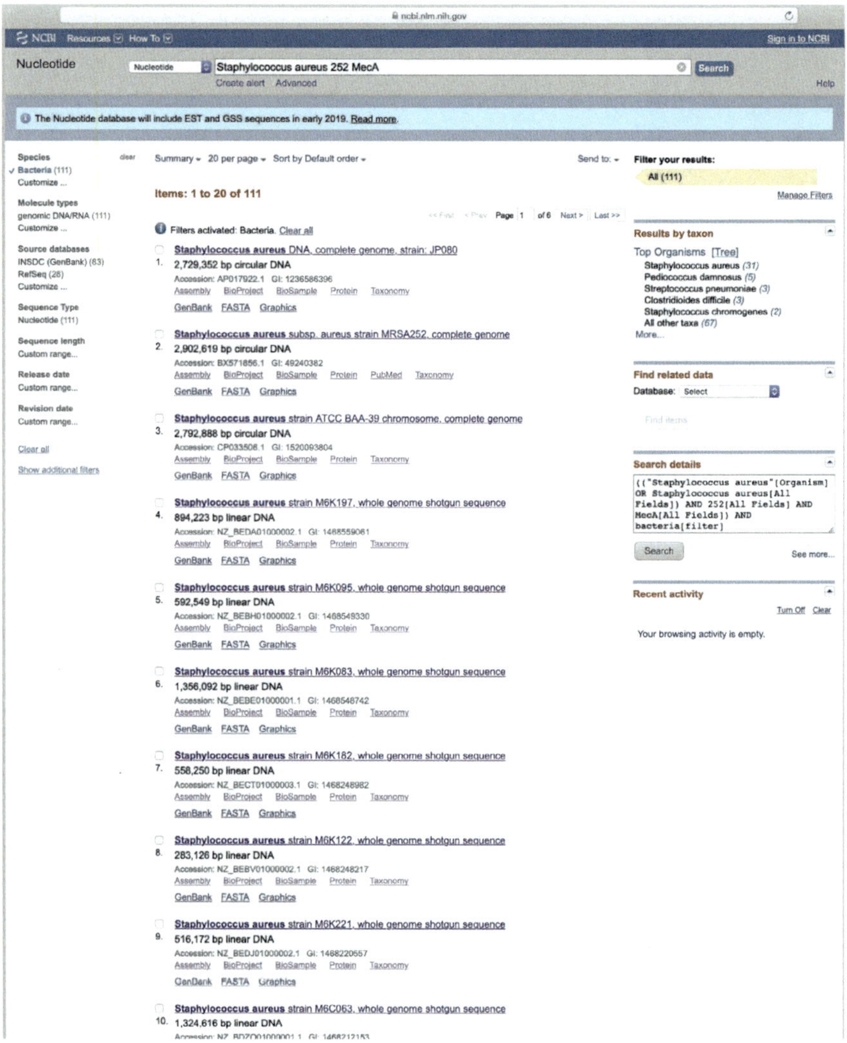

Figure 7.6 Nucleotide database search results for the search terms '*Staphylococcus aureus* 252 MecA'.

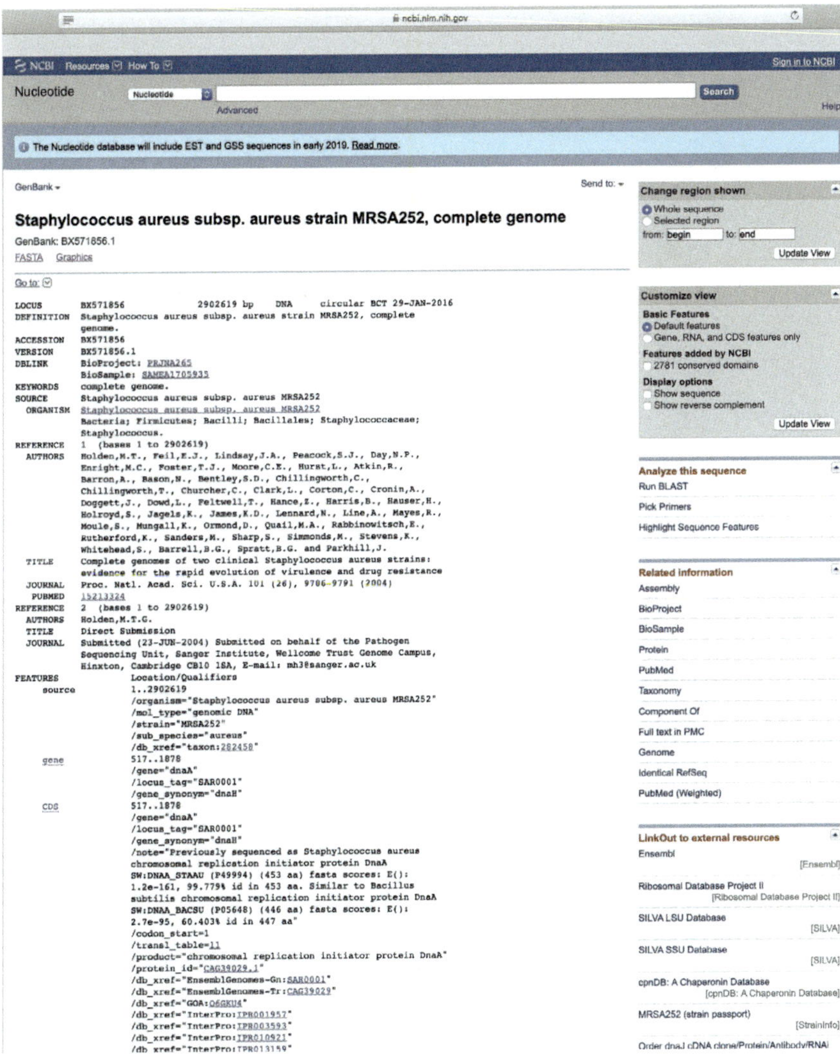

Figure 7.7 Gene information for the various genes found with the *Staphylococcus aureus* 252 genome.

Figure 7.8 The gene of interest, MecA, has been located by the Find function. In this example, the location of the gene name has been highlighted in *yellow* and the gene link that should be selected next to generate the nucleotide code is highlighted in *blue*.

Figure 7.9 The MecA gene-specific information includes the genetic code, which can be used as the template for primer design. This is found at the bottom of the page under Origin. The code can also be found by selecting CDS (Coding DNA Sequence) instead of Gene and then selecting FASTA. The code can then be copied and used manually to generate primers or it can be pasted into a primer design website such as primer BLAST on the NCBI website (Fig. 7.10) or a primer design site offered by multiple manufacturers, such as the one provided by Eurofins Scientific (www.eurofins.co.uk). These types of sites tend to be fairly straightforward to use.

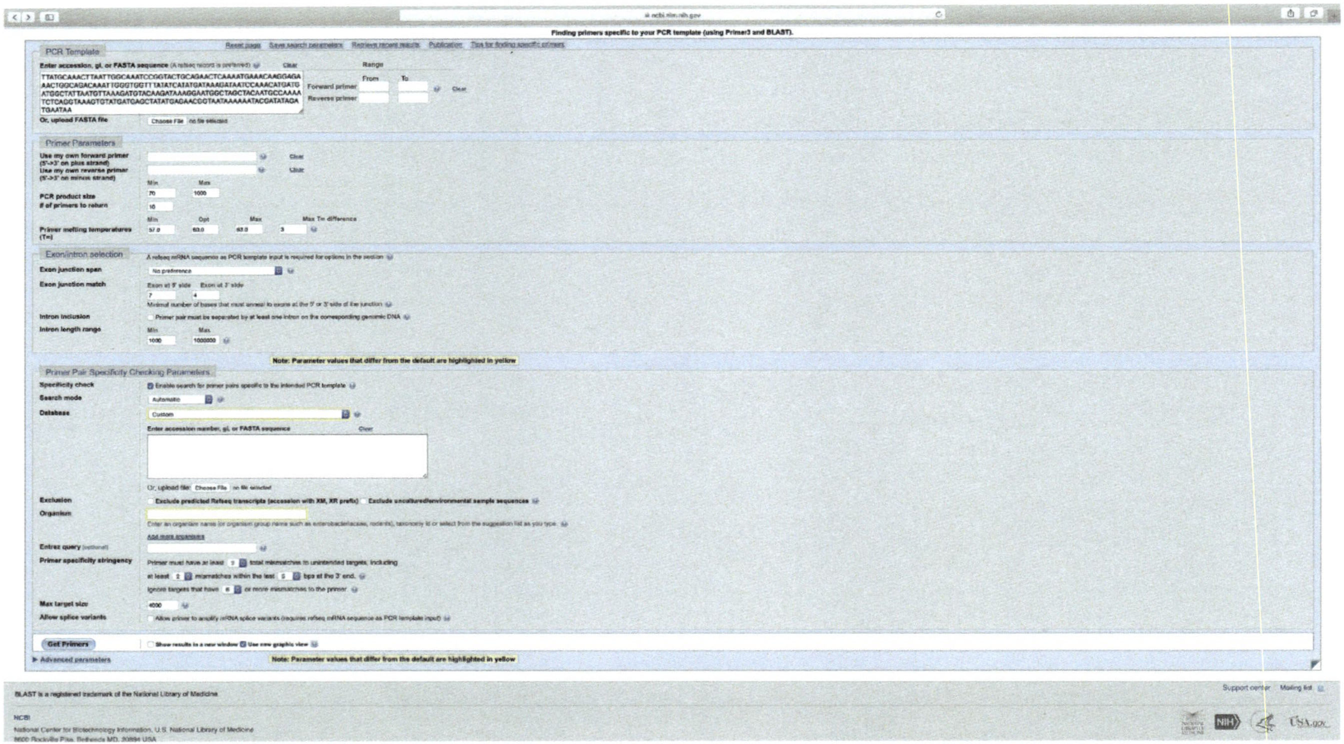

Figure 7.10 Primer design tool from the National Centre for Biotechnology Information website. The nucleotide code for the gene of interest has been entered into the template box. Below this are several other parameters that can be altered depending on the needs of the experiment.

Keep these points in mind when undertaking primer design. This will provide the best chance of creating effective primers. If using a design programme, once all parameters are set and submitted, the programme will usually suggest several sets of primers that would be suitable for the sequence suggested (Fig. 7.11). Sometimes it is impossible to fulfil all of these conditions when designing primers. Use the rules as a guide to optimise the likelihood of a successful PCR.

At the end of this process, you should have several sequences for primers that could amplify the gene of interest. There are multiple ways of getting to this point and lots of online sites that can help with finding the template nucleotide code and primer design. The ones suggested here may provide a starting point for this process if you have not carried it out before. It is always worth exploring the other available options. As suggested, it is also always worth checking published studies to see whether anyone has already designed the primer pairs needed for your reaction. If they have, it could save a lot of time.

Whichever way the primer sequences are decided upon, they will then need to be ordered. Primers can be purchased from a variety of companies such as Eurofins Scientific (www.Eurofins.co.uk) and Merck (www.sigmaaldrich.com). The most important step in this ordering process is ensuring that the primer sequence is entered into the ordering form correctly. Double-check or ask someone to read over the order before it is sent. Moreover, when the primers arrive, it is good practice to check primer sequences that were delivered before starting to use the primers. If the code of the primers which were delivered is incorrect and not checked, valuable time and money would be wasted in running PCRs with primers that will not work.

To reconstitute the primers, centrifuge the vial for a minute to ensure primers are at the bottom (the vial will look empty). Once centrifuged, add molecular-grade water to create a stock solution of 100 pmol/mL mix gently, aliquot and store.

To ensure the new primers are working and that the conditions under which they are working are optimal, a temperature gradient can be carried out on the first PCR in which they are used. Check whether the PCR machine in the laboratory has a temperature gradient facility on it. Using a range of temperatures from 5°C below and above the annealing temperature should be sufficient (the annealing temperature is roughly 5°C lower than then melting temperature). If running a PCR with new primers, it is also a good opportunity to optimise the primer concentration. A range of

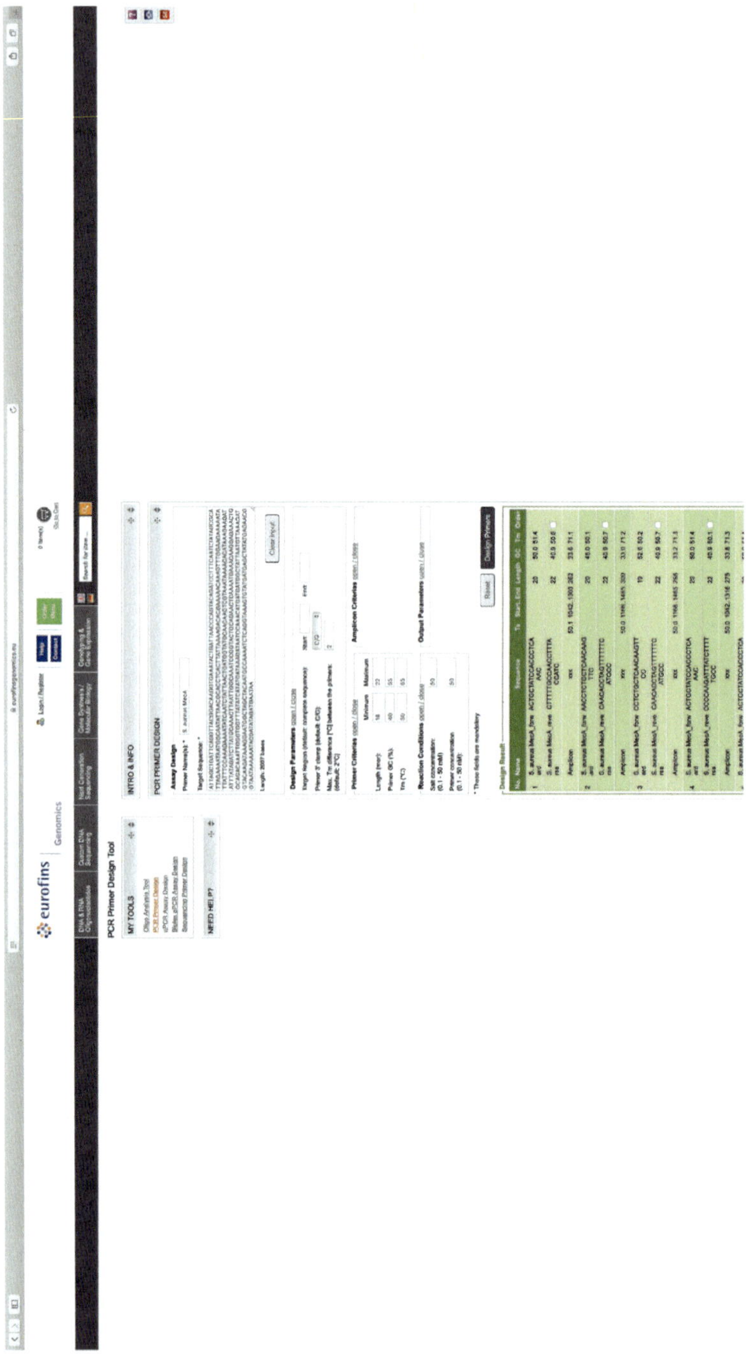

Figure 7.11 Selection of primers that can be ordered to detect MecA in *Staphylococcus aureus*. The predicted size of the product, melting temperature and guanine-cytosine content of primers are provided.

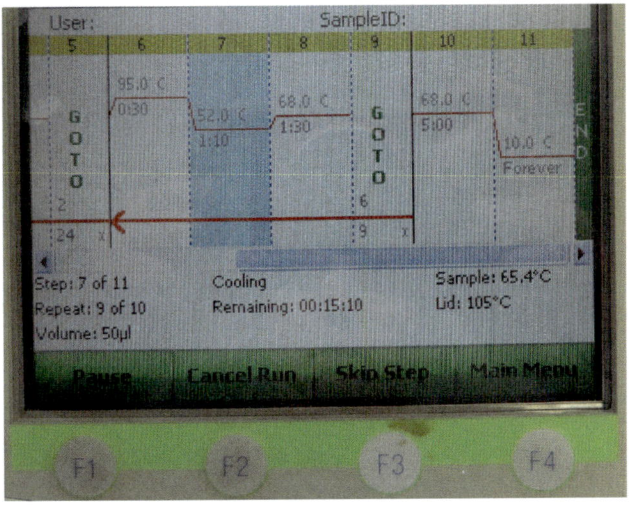

Figure 7.12 Polymerase chain reaction machine set up to run a thermal cycling profile. Many machines will display a profile on the screen along with the time it has been running and at what stage it is currently running.

concentrations between 0.1 and 1 μM should provide enough variation to find the best concentration for your reactions. Set up the PCR mixture in PCR tubes or plates as outlined in Section 7.2 and place it into the PCR machine. A thermal cycling profile will need to be programmed into the machine before it will run (Fig. 7.12). When setting up the thermal cycling profile, if the option is available to heat the lid of the machine, use it. It stops the reaction mixture from condensing on the lids of the tubes of plates. A general cycling profile that can be adapted for specific reactions is:

1. 95°C for 5 min: initial denaturation (once only)
2. 95°C for 30–60 s: denaturation
3. X°C for 30–45 s: annealing step (this needs to be calculated based on primers)
4. 70°C for 1–2 min: extension
5. 70°C for 8–10 min: final extension (once only)

Repeat Steps 2–4 for 25–40 cycles. If you are in the laboratory when the PCR cycle ends, use product straightaway as described in Section 7.4; if you not available when the PCR cycle ends (e.g. if you have left it to run overnight), set the machine to hold samples at 4–12°C indefinitely until they can be removed from the machine.

If the samples will not be used immediately once they are removed from the machine, store them at −20°C until needed. As with all experiments, it is important to run controls. For the negative control, replace the template

DNA with nuclease water. This will help identify whether the samples have been contaminated. For a positive control template, DNA can be purchased from a supplier, or if someone within the laboratory has a sample of template DNA known to work with the primers, this could be used.

7.4 Electrophoresis to analyse polymerase chain reaction products

The quickest way to determine whether a PCR has worked is to analyse it by agarose gel electrophoresis. Agarose is derived from seaweed, and when melted in an appropriate buffer, it polymerises to form a matrix through which DNA can move when an electric current is applied. Agarose gels can be of various concentrations. The concentration of the agarose determines the size of the pore of the gel matrix, and hence the size of DNA that can move through the gel. Agarose is dissolved in a buffer consisting of slats that allow an electrical current to travel. The most commonly used buffers for electrophoresis are *tris*-acetate-ethylenediaminetetraacetate (TAE) and *tris*-borate-ethylenediaminetetraacetate (TBE). Recipes for these buffers are given in Table 7.1.

To determine the concentration of agarose to use, it is important to know the size of the DNA fragment being analysed. Table 7.2 shows different agarose gel concentrations and the sizes of DNA fragment they are best suited to resolve.

To cast an agarose gel, you will need a casting tray (sealed at each end with tape) and a plastic comb that will make wells in the gel, into which you will transfer the DNA samples to be analysed. Casting equipment can be

Table 7.1 Ingredients to make working and stock solutions of *tris*-acetate-ethylenediaminetetraacetate (TAE) and *tris*-borate-ethylenediaminetetraacetate (TBE) buffers for electrophoresis.

Buffer	Working solutions	Stock solutions
TAE	1×	50×
Tris-acetate-EDTA	40 mM *tris*-acetate	242 g Tris base
	1 mM EDTA	57.1 mL glacial acetic acid
		100 mL 0.5 M EDTA (pH 8)
TBE	0.5×	5×
Tris-borate-EDTA	45 mM *tris*-acetate	54 g Tris base
	1 mM EDTA	27.5 mL boric acid
		20 mL 0.5 M EDTA (pH 8)

Stock buffers can be prepared, stored at room temperature and diluted to the working concentration when needed. *EDTA*, ethylenediaminetetraacetic acid.

purchased from numerous suppliers. Each will tell you the typical volume of gel that the casting tray can hold: for example, it might be 50 mL (Fig. 7.13).

Therefore, to cast a 0.8% (w/v) agarose gel in a 50-mL tray, you will need to weigh 4 g of agarose and mix it into 50 mL TAE or TBE buffer. The agarose will dissolve only when heated, so either heat the mixture using a Bunsen burner and tripod until the buffer begins to bubble and the agarose dissolves or microwave it at a low to medium temperature for 1-min intervals with mixing (swirling) after each minute. Wear personal protective equipment when doing this. Once the agarose has dissolved, allow the liquid gel to cool a little (1–2 min) or place it in a 50°C water-bath until you are ready to cast the gel. Then, pour the liquid gel into the casting tray

Table 7.2 Concentrations of agarose that will resolve different-sized DNA fragments.

% (w/v) Agarose	Optimum resolution (kb)
0.5	1–30
0.7	0.8–12
1.0	0.5–10
1.2	0.4–7
1.5	0.2–3
2.0	0.05–2

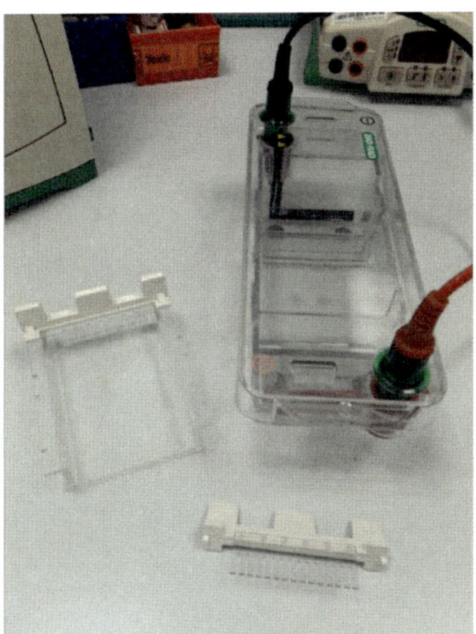

Figure 7.13 Gel casting tray, comb and electrophoresis tank.

and allow it to set (Fig. 7.14). When the gel is set, it should appear slightly opaque and white (Fig. 7.15). It is possible to add a DNA dye to the piqued gel before casting it (see later discussion).

Once the gel is completely set, gently but firmly remove the comb. Where the teeth of the comb were, there will be a series of well into which

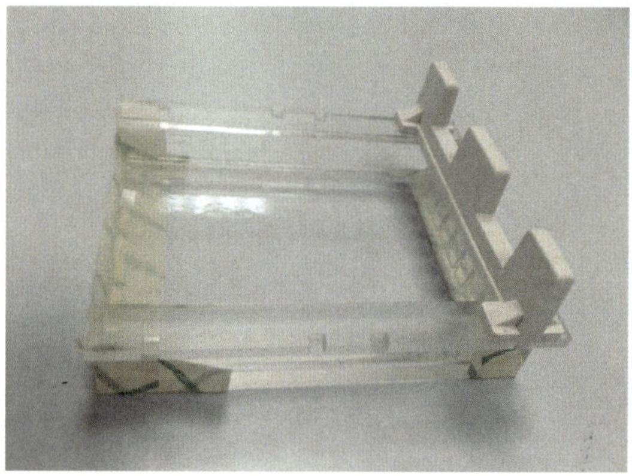

Figure 7.14 Agarose gel cast in a gel-casting tray that has been sealed at each end with tape. The comb is present; when removed, it will reveal a number of wells. The set gel has a slightly opaque appearance.

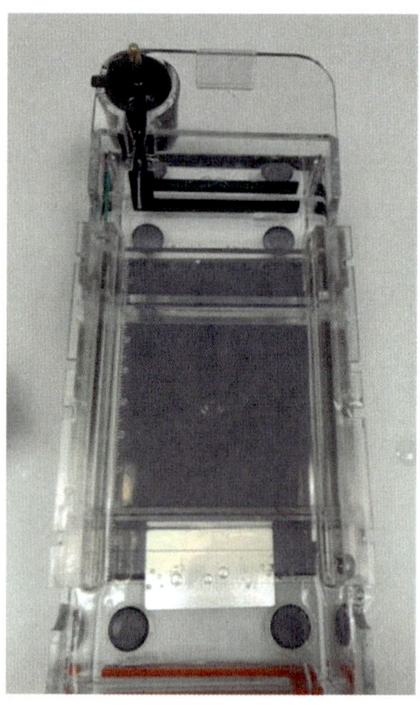

Figure 7.15 Gel loaded into the gel tank and covered with electrophoresis buffer. A black card has been placed underneath the gel so that the wells are visible. The wells are closest to the cathode (*black*).

you will transfer the DNA samples/PCR products. Make sure the number of wells is sufficient for the number of samples being analysed, with two spare wells for a molecular weight marker. If the gel-casting tray has been sealed with tape, remove and discard it.

Place the gel into the electrophoresis module. DNA is negatively charged so when an electric current is applied to the gel tank the DNA will move towards the anode, the anode is usually colour coded red. When you place the gel into the electrophoresis module, make sure the wells are closest to the cathode (colour-coded black) so that the DNA will move through the gel in the correct direction (Fig. 7.15). Pour enough TAE or TBE (use the same buffer that you used to make the gel) to cover the gel to a depth of approximately 2 mm, ensuring that all of the wells are filled. If there are air bubbles in the wells, remove them with a pipette tip.

A molecular weight marker must be used when analysing PCR products by electrophoresis. Choose a molecular weight marker that has a series of fragments that include the size of the PCR product you are analysing. For example, if your PCR product is 1 kb, a molecular weight marker ranging from 200 bp to 10 kb resolved on a 0.8% agarose gel is ideal. Molecular weight markers usually come pre-prepared in a coloured loading dye; if not, follow the manufacturer's instructions to obtain the correct dilution in the correct volume of coloured loading dye. The coloured loading dye helps you to see the sample when it is being pipetted into the well. It also contains glycerol, sucrose or Ficoll, which makes the sample sink into the well and not float away. In addition, the coloured dye helps you to see how far the PCR samples have moved through the gel, because the dye will also move toward the anode ahead of the PCR product (which will not be visible). Before the PCR product is loaded into the well, it must be mixed with a coloured loading dye, too. Table 7.3 gives some common recipes for loading dyes.

The loading dye must be diluted to a working concentration (1×), which is achieved by mixing with the PCR product. For example, when loading 5 μLL of a PCR product onto a gel in 6× loading buffer you would add 1 μL of loading buffer to 5 μL of the sample (and mix) and transfer 5 μL into the well. These volumes dilute the 6× loading buffer to the 1× concentration required. The samples are then ready to load into the wells of the gel. Using a fine pipette tip (10 μL maximum volume), transfer each sample to a separate well, including the molecular weight marker. The molecular weight marker should go into the first and the last wells, with the samples to be analysed in between (Fig. 7.16). When doing this, place the pipette halfway into the well and gently depress the plunger to allow the sample to sink

Table 7.3 Recipes for four different types of loading dye. L **Loading dye at 6× concentration**

Bromophenol blue: 0.25% (w/v)
Sucrose: 40% w/v (water)
Bromophenol blue: 0.25% (w/v)
Xylene cyanol FF: 0.25% (v/v)
Sucrose: 40% w/v (water)
Bromophenol blue: 0.25% (w/v)
Xylene cyanol FF: 0.25% (v/v)
Glycerol: 30% w/v (water)
Orange G: 0.025% (w/v)
Xylene cyanol FF: 0.25% (v/v)
Sucrose: 40% (w/v) water

These can be made in 10–20mL volumes and aliquoted into smaller portions for long-term storage (refrigerated or frozen). For each of these, sucrose (40%), glycerol (30%) and Ficoll (15%) can be alternatively used.

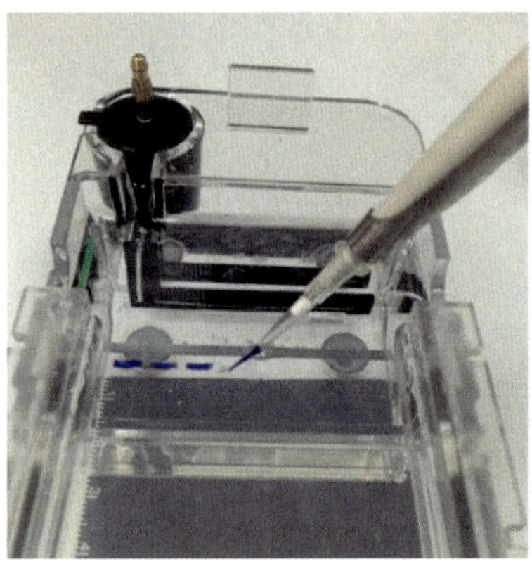

Figure 7.16 Gel being loaded with DNA samples mixed with blue loading dye.

into the well. Do not put the tip to the bottom of the well because this will damage the agarose. Do not pipette too quickly or the turbulence created will make the sample float out of the well.

Once all of the samples are in the well, place the lid onto the electrophoresis module (aligning the colour-coded anode and cathode) and apply an electric current. The manufacturers of electrophoresis modules will also sell power packs designed for use with them, so ensure you use the right one set to the recommended voltage. As a rough guide, it is possible to resolve a 1-kb PCR product on a 0.8% agarose gel at 90V for 45min, at which point you will observe that the dye front will have moved just over two-thirds of the way down the gel.

DNA is colourless, so to visualise the PCR product, it must be dyed. The most commonly used DNA dye is ethidium bromide, which can be visualised under UV light. This chemical is carcinogenic and also a known mutagen, so care must be taken when handling it, including using protective gloves and eye protectors to avoid splashes. Ethidium bromide binds nonspecifically to double-stranded DNA (dsDNA) by intercalating with it. DNA can be dyed after electrophoresis is completed by transferring the gel to a container of TAE or TBE buffer (enough to cover the gel) containing 0.5 $\mu g\,mL^{-1}$ ethidium bromide. The gel is left immersed in the staining solution for 30–45 min and then is destained by rinsing in water to reduce background staining. The ethidium bromide-containing buffers must be discarded into a sealed container and deactivated with charcoal before disposal (see your local laboratory guidelines). Alternatively, ethidium bromide can be added to the liquid gel before casting, which means that the DNA is dyed during electrophoresis. The TAE/TBE buffer used for electrophoresis must be treated before disposal as described earlier. Alternatives to ethidium bromide are available, such as SYBR Safe, which should be used according to the manufacturer's instructions.

To visualise ethidium bromide-stained gels, a UV transilluminator is used. These consist of a tray onto which the gel is placed (without the casting tray). The UV light source is located beneath this, and so the gel is illuminated from underneath. UV transilluminators are housed inside a box, with a door or drawer into which the gel is placed. This is because the UV-illuminated gel must be observed in the dark; otherwise, it is difficult to see the stained DNA. The UV transilluminator is ordinarily connected to a computer and the image of the gel will appear on the screen. The gel will appear black with stained DNA as white bands (Fig. 7.17). Use your knowledge of the expected size of the PCR product and the known sizes of the molecular weight marker to decide whether the PCR has been successful.

A successful PCR should produce a single band of the correct molecular weight. Additional bands indicate nonspecific primer binding, very low-molecular weight bands show that the primers have dimerised (which may not impair the success of the PCR) and no bands indicate that the PCR has not worked. There are a number of reasons for this, which will be described in Section 7.6. Having visualised the gel, make sure you save or print an image of the results before disposing of the gel according to local laboratory guidelines.

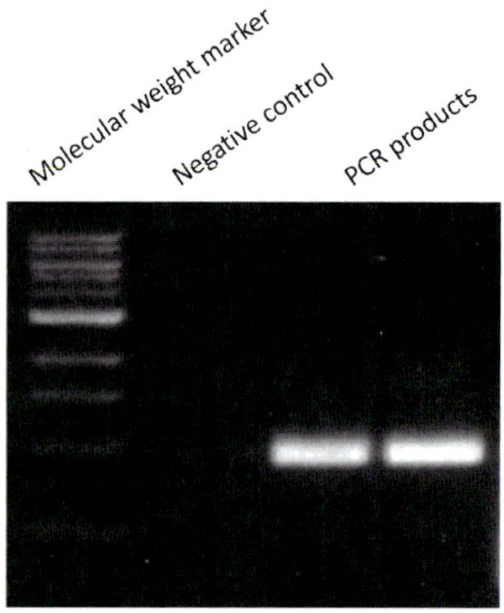

Figure 7.17 Gel that has been stained with ethidium bromide. The ethidium stains the DNA bands, which appear white against a black background when photographed using a UV transilluminator. *PCR*, polymerase chain reaction.

7.5 Quantitative polymerase chain reaction

In addition to detecting the presence of a particular gene or DNA fragment, PCR can be used to assess the expression levels of a gene, to indicate how an organism alters gene expression in response to environmental changes. End-point PCR, which relies on electrophoresis to observe a gene product, can be employed to do this. PCRs are broadly prepared as described in this chapter, except the template for the PCR is cyclic DNA (cDNA) rather than genomic DNA.

For this type of PCR, instead of extracting total genomic DNA from the organisms of interest, the total RNA is extracted. The easiest way to do this is to use a commercially available RNA extraction kit or the TRIzol method:

1. Pellet bacterial cells by centrifuging at 13,000 rpm for 5 min in a microcentrifuge tube.
2. Add 1 mL TRIzol and incubate at room temperature for 5 min.
3. Add 250 μL chloroform and shake vigorously; incubate at room temperature for 5 min.
4. Pellet the cell debris by centrifuging at 13,000 rpm for 5 min.

5. Three layers will be visible in the tube: the top layer will be clear (aqueous phase); the middle layer will be white (precipitated DNA); the bottom layer will be pink (organic phase).
6. Using a pipette, carefully transfer the top phase to a fresh microcentrifuge tube; leave some behind (approximately 1 mm deep) to avoid accidently collecting the precipitated DNA.
7. Add 550 μL isopropanol to the fresh tube containing the aqueous layer and mix gently by pipette; incubate at room temperature for 5 min.
8. Centrifuge at 13,000 rpm for 30 min.
9. Place the tube on ice, decant the isopropanol and add 1 mL 75% ethanol (prepared in RNase-free water). Mix gently by pipette and centrifuge at 10,000 rpm for 5 min.
10. Decant the ethanol and gently tap the tube on a paper towel to remove any residual drops from the bottom of the tube.
11. Leave to air dry.
12. Add 25–50 μL RNase-free water to the tube and mix by gently flicking.
13. Store RNA at −80°C

The RNA must be quantified and checked for purity, which is done by spectrophotometry using a quartz cuvette and two readings taken at 260 and 280 nm, or it can be done using a Nanodrop, which works on the same principle but does not use a cuvette. For DNA, the 260/280 ratio should be 0.8, and for RNA it should be 2. Various factors can result in a low ratio, including ethylenediaminetetraacetic acid, phenol or proteins.

To calculate the concentration of DNA/RNA based on the 260 or 280 readings, perform the following calculation, bearing in mind these two points:

The optical density (OD) at 260 nm (OD_{260}) equals 1.0 for the following solutions:

a 50-μg/mL solution of dsDNA
a 40-μg/mL solution of RNA

Example calculation:

A sample of dsDNA was diluted 10×.

The diluted sample gave a reading of 0.25 on a spectrophotometer at OD_{260}.

dsDNA concentration = 50 μg/mL × OD_{260} × dilution factor
dsDNA concentration = 50 μg/mL × 0.25 × 10
dsDNA concentration = 6250 μg/mL or 0.62 mg/mL

RNA is very unstable, so ensure you work in a clean area using molecular-grade, RNase-free reagents, and wear gloves when handling all samples and containers. If possible, section off a corner of your bench (or laboratory) and use a designated set of pipettes for RNA work only. It is advisable to use filter tips when pipetting.

Next, cDNA must be reverse-transcribed from the total RNA. To do this, a commercially available kit must be used according to the manufacturer's instructions. Some kits rely on a separate step before PCR; others are provided as a combined reverse-transcription/PCR kit. Make sure that equimolar quantities of RNA are used in the PCRs so that any changes in gene expression seen are the results of a real effect and not because you added different amounts. In addition to the target gene you are analysing, make sure you include at least one housekeeping gene (HKG) whose expression does not change in response to the environmental conditions tested. A commonly used HKG is GAPDH, but it is worthwhile to conduct a quick search of the literature to see what HKGs are regularly used to study your bacterium of interest.

For end-point PCR, in which the PCR product is analysed by gel electrophoresis, standard PCR reagents are used. The product is then analysed by agarose gel electrophoresis, as described earlier. The amount of PCR product is quantified by densitometry, which is a measure of the relative brightness of the band, and so is really only semi-quantitative. Because of this, it is imperative to ensure that exactly the same volume of each PCR product is loaded onto the gel to ensure that any differences are not the result of inaccurate pipetting. Once the gel has run, it is photographed in the same way as described previously. The brightness of the bands is determined using software usually provided with the transilluminator, which measures the brightness of the bands arbitrarily. The number allocated to each band can be used to determine relative increases and decreases in gene expression. For example, if the brightness is higher under the condition tested, the gene expression is increased; if it is lower, the gene expression is decreased. Test bands under a given condition must be compared with a standard (e.g. untreated versus treated) condition. The results are valid only if there is no change in the expression of the HKG under the same two conditions. Any differences in expression can be presented as a percent increase or decrease. It is possible to calculate molar values for each band if the molecular weight marker used is a quantitative marker, in which the size and amount of DNA in each band are known. The principle of the comparison is the same, but using the molecular weight marker, you can estimate how much DNA is in the sample. For example, if the band of the marker has a densitometry value of 2500 and there is 100 ng of DNA in the band, and the test sample has a densitometry value of 1,250, you can estimate that there is 50 ng of DNA in the test sample.

Gene expression can also be quantitatively analysed in real time. This principle is the same, but the data are not analysed by electrophoresis, and you are able to see the PCR product accumulate over time rather than observing a snapshot at the end of the reaction. For this type of PCR, cDNA is used as the template, as previously described, and the PCR is prepared in the same manner as usual. The difference is that a fluorescent DNA dye is added to the reaction. The most commonly used dye is SYBR green, which binds to dsDNA. This means that as the PCR product accumulates, SYBR green binds to it and produces a fluorescent signal. The more product accumulates, the higher these signal, until the reagents become a limiting factor. A real-time PCR machine is needed for quantitative PCR because these have the capacity to detect the fluorescent signals, whereas a standard PCR machine does not. The fluorescent signal is read and interpreted by a computer and presented as an exponential curve in real time (Fig. 7.18). A negative control must also be included; it contains no template and so it should not produce a curve as the reaction proceeds. If it does, it suggests that there is a contaminant in the reaction.

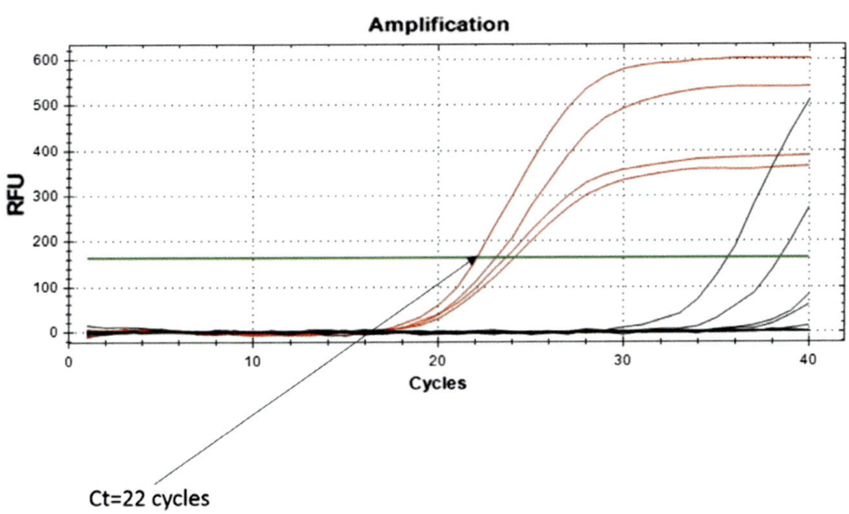

Figure 7.18 Typical result output for quantitative polymerase chain reaction (PCR). The relative fluorescent units (RFU) are seen on the y axis and indicate the accumulation of the PCR product. The x axis is the number of PCR cycles. The cycle threshold (Ct) value for one of the reactions is highlighted and is the point at which the trace crosses the threshold line, in this case after 22 cycles.

HKGs must be analysed at the same time as the test gene and their expression should not change under the conditions tested. The threshold for detecting the gene of interest must also be established. Determining the threshold of detection will require a series of experiments in which the template material is titrated to form serial dilution. Primers must be similarly titrated to ensure optimal binding to the target and avoid spurious amplification, or incomplete amplification caused by the primers forming dimers, all of which can lead to a high fluorescent background. These experiments must be done before the gene expression analysis.

The simplest way to interpret gene expression from quantitative PCR data is to calculate the fold change. This can be done using the comparative cycle threshold (Ct or ΔΔCt). The Ct value is the number of PCR cycles needed for the fluorescent signal to cross the threshold (previously determined by titration of template and primers to exclude any background noise, described earlier). Ct values are inversely proportional to the amount of target nucleic acid in the sample (the lower the Ct level, the greater the amount of target nucleic acid in the sample).

The ΔΔCt method compares the Ct value of one target gene to another (the HKG or reference gene). The following points and Table 7.4 set out how to calculate ΔΔCt, but most software packages provided with real-time PCR machines will do this and the information can be exported to programs such as Excel:

1. Calculate the average Ct values for the HKG under the test conditions (treated) and under control conditions (untreated), and the gene of interest (GOI) under the test and control conditions. This will give you four values: HKG treated, HKG untreated, GOI treated and GOI untreated.
2. For the treated samples, subtract the average Ct for HKG from GOI; this is the ΔCt treated.
3. For the untreated samples, subtract the average Ct for HKG from GOI; this is the ΔCt untreated.

Table 7.4 How to tabulate experimentally determined cycle threshold (Ct) values and use this information to determine the fold change in gene expression.

Treated		**Untreated**					**Fold change**
Ct for GOI	**Ct for HKG**	**Ct for GOI**	**Ct for HKG**	**ΔCt treated**	**ΔCt untreated**	**ΔΔCt**	**(Log_2ΔΔCt)**
23.47	21.53	21.12	20.94	1.94	0.18	1.76	3.38

GOI, gene of interest; *HKG*, housekeeping gene.

In this example, the fold change indicates that the gene expression is increased. If the value were −3.38, it would indicate that the expression decreased.

4. Next, subtract the ΔCt untreated from the ΔCt treated; this gives you the ΔΔCt value.
5. To convert this to a fold change, calculate the log base 2 value of the ΔΔCt.

7.6 Troubleshooting

Successful PCR is a useful and powerful tool; however, it can be tricky to optimise a new reaction. Table 7.5 shows common problems encountered when optimising a PCR, with some possible causes and solutions.

Table 7.5 Common problems that might be encountered when optimising a polymerase chain reaction (PCR), and possible solutions.

Problem	Possible causes and solutions
No product	• insufficient template added. (Try extraction again; quantify amount if in doubt about yield; try a range of template concentrations) • primer not added (ensure you spot small volumes onto inside of tube so you can see you have added them; mix contents thoroughly before adding to PCR machine). • deoxyribonucleotide triphosphates not added (see previous suggestion) • master mix concentrations wrong (check ratios of ingredients to ensure they are correct) • impure template/inhibitors (extract again or heat sample to 95°C for 5 min before use in PCR machine) • thermal cycler not functioning properly (test again with known primers to check)
Background amplification	• too much template added to reaction (reduce amount of template added until background is reduced) • too many cycles were used (reduce number of cycles used) • too much primer (reduce amount of primer used; can use titration) • improper annealing temperature (check annealing temperature of primers; set gradient of temperatures in 1°C steps) • poor-quality template (isolate new template) • poorly designed primers (check that there is no secondary structure or complementarity between primers; redesign if necessary)
Nonspecific amplification	• primers hybridising to more than one site on template (increase annealing temperature by gradient or design new primers) • contamination! Make new stocks of primers/buffers/dna template

7.7 Notes page

Record observations and notes here.

CHAPTER EIGHT

Screening for common virulence traits

Contents

8.1 Introduction

Bacteria can be divided into categories depending on their ability to cause an infection. These are broadly known as commensals (do not cause infection) and pathogens (can cause infection). Bacteria that can cause disease are referred to as pathogens or opportunistic pathogens, and the severity of the disease they cause is determined by their virulence. Bacteria can express different virulence factors that help them attach to and invade host cells, evade the immune system, scavenge for nutrients or produce toxins, to name a few. A series of basic tests can identify which virulence factors uncharacterised bacteria might be capable of producing. This chapter will familiarise you with some tests you can use to determine the virulence characteristics of a bacterium of interest.

8.1.1 Agar-based tests

There are several different ways in which agar plates can be set up to test for virulence, including streaking or stabbing the agar with an inoculum and creating wells for a bacterial culture or supernatant. Some of the various methods are outlined below. Fig. 8.1 shows some of the equipment needed to set up the agar-based methodologies.

Haemolysin activity - Haemolysins are enzymes secreted by many different bacteria. They act by damaging the cell membrane and lysing host red blood cells. The extent to which bacteria break down blood cells

Bacteriology Methods for the Study of Infectious Diseases
ISBN 978-0-12-815222-5
https://doi.org/10.1016/B978-0-12-815222-5.00008-0

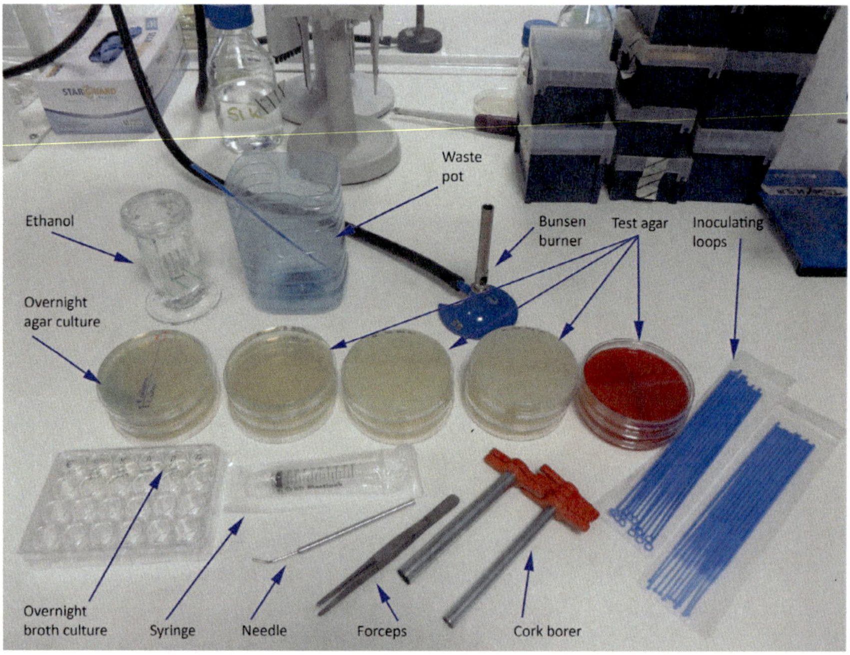

Figure 8.1 Basic set of equipment that might be needed to set up agar plates to test for virulence, including different agars (egg, milk, DNase and blood), inoculating loops, bacterial cultures and sterilising equipment.

depends on what type of haemolysin they produce. There are three main types of haemolytic activity:

1. **α-haemolysis:** The haemolysin released by the bacteria will partially break down the host blood cells but does not completely break down the host haemoglobin;
2. **β-haemolysis:** The haemolysin released by the bacteria completely breaks down the host blood cells and haemoglobin; and
3. **γ-haemolysis:** The bacteria produces no enzymes with the ability to break down the host blood cells.

Knowing the haemolytic profile of an unknown isolate can help with identifying that isolate in conjunction with other biochemical tests. Knowing the baseline haemolytic profile of bacterial isolates also allows the opportunity to look at any change in haemolytic activity caused by a change in conditions (e.g. treatment with a novel antimicrobial agent or a change in incubation temperature).

Screening isolates for activity is fairly simple and requires only the bacterial sample, blood agar plates, inoculating loops, a Bunsen burner, a waste pot and an incubator. If you are making your own plates, you will also need a 5-mL pipette, sterile blood and sterile Petri dishes into which to pour the agar.

Blood plates can be bought premade from suppliers such as Thermo Fisher Scientific (www.thermofisher.com) or made up in the laboratory. If purchasing premade plates or blood, keep in mind that different types of blood are available (e.g. horse, sheep and rabbit) and check which type of blood is most suitable for the type of experiment or bacteria you are running. Horse blood plates are commonly employed, but you can get different haemolytic reactions depending on which blood is used. Once the blood plates have been delivered, store them in the refrigerator at 2–8°C and note the expiration date.

When making up your own blood plates, order in blood to add to the agar base that you choose to use. If there is a choice, it is advisable to get defibrinated blood because this ensures the blood will not clot when you are working with it. Defibrinated horse blood is the best choice when testing nicotinamide adenine dinucleotide-requiring organisms, because sheep blood does not support their growth in the same way. Store the blood in the refrigerator and make sure it is used before the expiration date. To make up blood plates, autoclave the agar that will be used. This can be a specific blood agar base; nonspecific agars are also used: for example, tryptic soy agar. Once it is autoclaved, move the bottle holding the agar to a water-bath set at 45–50°C. While the agar is cooling, remove the blood from the refrigerator and allow it to come to room temperature. It is important to allow the agar to cool sufficiently so as not to damage the blood when it is added. it is also important not to add refrigerated cold blood to the agar, because if the agar cools too much, it can start to set and become lumpy before it has been poured into the Petri dishes.

When the agar is cool enough, add 5% (v/v) blood. To do this, use either a serological pipette or an adjustable pipette. Take the required volume of blood from its sterile container, flame the neck of the bottle containing the cooled molten agar, add blood to the agar, flame the neck of the bottle again and replace the lid. Make sure the work is carried out aseptically. Swirl the bottle containing the agar and blood to mix them together thoroughly before you start to pour. Do not shake the bottle because it will create bubbles that will sit on top of the agar plate when you pour them. To pour the plates, stack the labelled Petri dishes close to the Bunsen burner and, starting at the bottom, pour 20 mL of the blood agar into each Petri dish. Leave the plates to set on the bench and do not move them once they have started to set, because this will disrupt the surface of the agar. Once made, it is best to use these plates within 7 days.

To test for haemolysin activity, take a single colony of bacteria from an overnight agar plate and inoculate it onto a blood plate (that has warmed to room temperature) as a short streak; then, incubate the plate overnight at about 37°C (Fig. 8.2).

The same procedure can be followed using a liquid bacterial culture. However, if the liquid culture is contaminated and more than one species is present, this may not become apparent until after the incubation. When using a broth culture, it is a good idea to standardise the number of bacteria being used, which makes it easier to compare samples. One way to standardise the bacterial culture would to be to follow a method similar to that used in antimicrobial susceptibility testing (see Chapter 4). Briefly, aseptically pick a few bacterial colonies from an overnight agar plate culture and suspend these in sterile saline to give a density equivalent to a 0.5 McFarland standard. Use this culture to inoculate the agar plate.

A second method for looking at haemolysin activity involves the same setup, but instead of streaking the bacteria onto the plate, a stab inoculating loop can be used to stab the blood agar at a single point (Fig. 8.3). The plate should then be incubated in the same way at about 37°C for 24 h.

A third method can be used to detect activity; this involves creating wells within the agar using a sterile metal cork borer of a known diameter. Ensure that the cork borer is sterile (this can be done by flaming in alcohol), remove the lid of the agar plate and push the cork borer into the agar. Remove the cork borer from the agar (the agar plug should come with it) and there will be a well in the plate. If the agar plug does not come away when the cork borer is removed, it can be taken out using sterile forceps (Fig. 8.4).

To test for haemolysin activity, pipette a known volume of bacterial culture and incubate the plate at about 37°C (do not invert the plates for incubation) for 24 h. The diameter of a zone of haemolysis around the well can then be measured. This method can be useful if you are planning on looking at changes in haemolysin production (e.g. caused by exposure to a subinhibitory antibiotic). The activity of haemolysin is affected by the growth of the bacteria while incubated, so in addition to standardising the bacterial culture before starting the method, you can adapt this to look at the extracellular activity of haemolysin. This is achieved by filtering the bacterial culture through a 0.22-μm filter and then filling the well with the filtered bacterial supernatant before incubating it and measuring it alongside plates that contain bacteria.

(A) (B) (C)

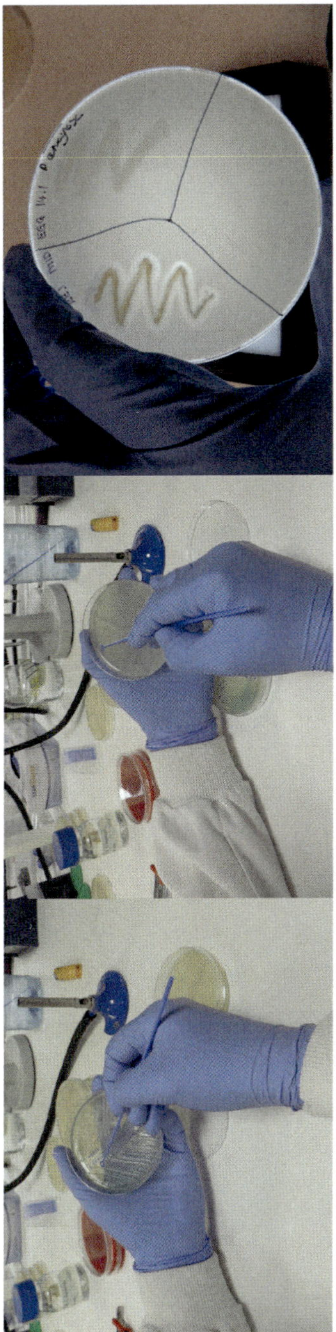

Figure 8.2 Universal example of how to streak bacteria onto an agar plate to investigate virulence activity (note that the agar being used is not blood agar). (A) Bacterial colonies are taken from an overnight culture. (B) Bacteria are inoculated onto a sterile agar plate. (C) Resulting growth and activity.

(A) (B) (C)

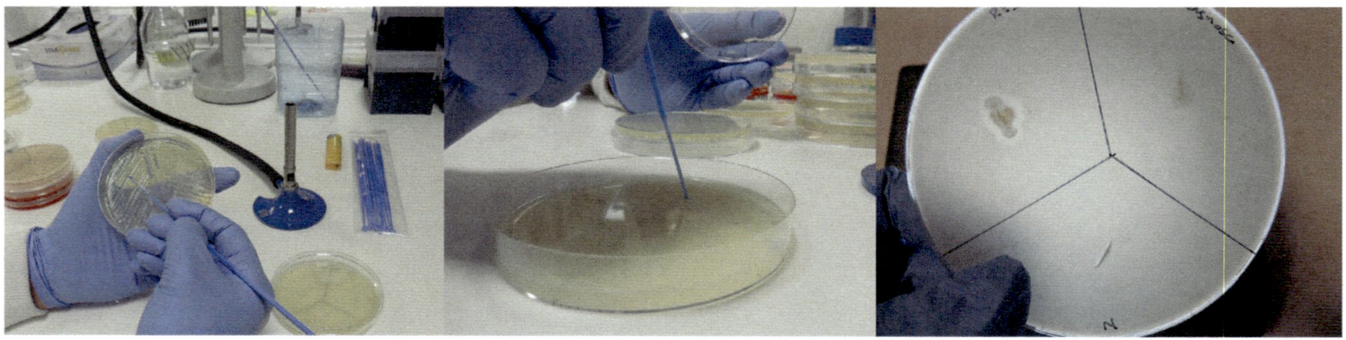

Figure 8.3 Universal example of how to create a bacterial stab agar plate to investigate virulence activity (note that the agar being used to demonstrate the technique is not blood agar). (A) Bacterial colonies taken from an overnight culture. (B) Bacteria inoculated or stabbed into a sterile agar plate. (C) Resulting growth and activity.

(A) (B)

(C) (D)

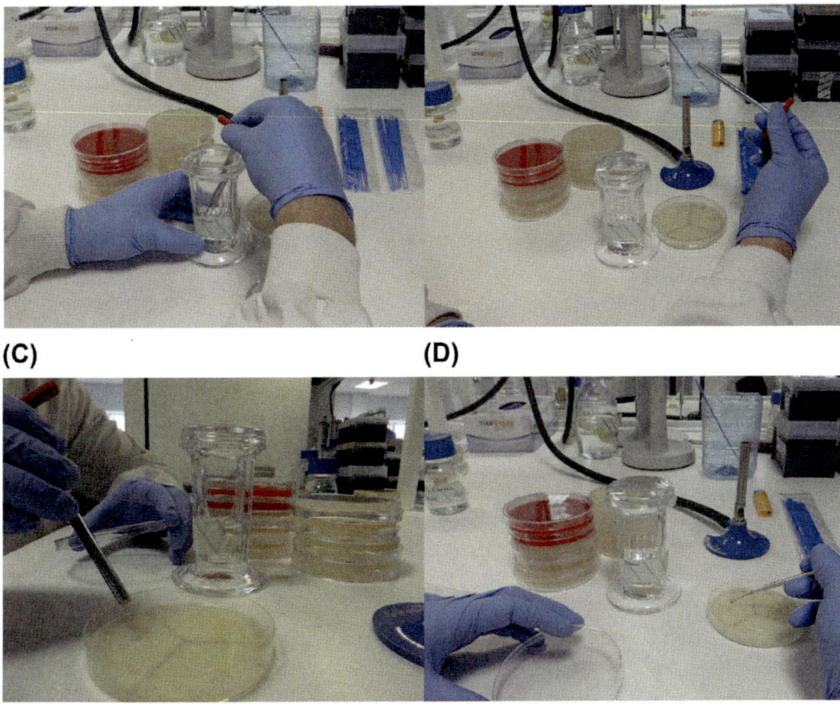

Figure 8.4 Universal example of how to create wells within an agar plate in preparation for testing the virulence of a bacterial culture or supernatant. (A) Cork borer is dipped into alcohol. (B) Cork borer is passed through a Bunsen flame to burn off alcohol (C) Cork borer is pressed into agar and removed, leaving a well. (D) If the agar plug does not come away with the cork borer, a needle can be used to remove the remaining agar.

After incubation, the plates commonly display one of three distinct reactions:

1. α-haemolysis: α-haemolysis will give a greenish colour in which the blood has been partially broken down;
2. β-haemolysis: β-haemolysis will give a clear zone on the agar where the bacteria have completely broken down the blood cells and haemoglobin; and
3. γ-haemolysis: γ-haemolysis will show bacterial growth with no effect on the agar itself (Fig. 8.5).

If it is unclear what reaction has taken place on the plate, it can help to put the plate onto a light box (as shown in Fig. 8.5) so that the reactions are clearer or to hold the plate over a piece of paper that has writing on it. It should not be possible to read the words through the blood plate or through

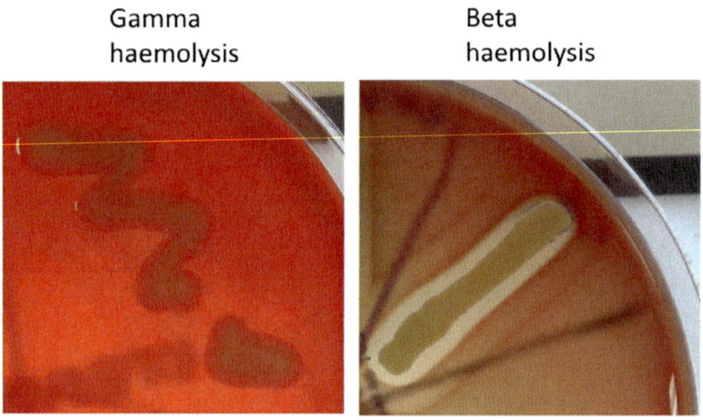

Figure 8.5 Horse blood agar plate (left) and sheep blood plate (right) after 24 h incubation at 37°C with bacteria streaked onto surface demonstrating γ- and β-haemolysis, respectively.

the α-haemolysis reaction, but if β-haemolysis has occurred, it should be possible to make out the words through clear zones around the bacterial growth.

If you are unsure what these reactions look like, it is a good idea to run controls. Un-inoculated blood can be incubated as a blood plate control. Positive controls for both α- and β-haemolysis can be run by using an isolate with a known haemolysin profile alongside the test organism, which allows a comparison of the activity of an unknown isolate with known activity.

Protease activity - The ability of bacteria to break down and process peptide bonds is an important function that affects the ability of bacteria to cause an infection. Different proteases can act on different proteins, affecting the ability of a bacterium to degrade and invade host tissue. It is possible to test bacterial isolates to determine whether they have protease activity by using casein hydrolysis. Casein is a large protein found in milk that must be broken down by bacteria into amino acids to be employed. Identifying the protease capabilities of an unknown organism can help with bacterial identification, or once the normal activity profile is known, changes in protease activity caused by changes in conditions can be investigated. To test for casein activity, you will need a bacterial isolate, agar, inoculating loops, a pipette, a Bunsen burner or hotplate, a waste pot, a tripod and gauze, a beaker, sterile Petri dishes and dehydrated milk powder or sterile milk.

Screening for protease activity, such as using the haemolysin test, is a fairly simple and rapid procedure that requires milk plates in place of blood plates. Milk plates and milk agar can be bought directly from a supplier in the same way as the blood plates. If purchasing premade milk plates from a supplier, follow the manufacturer's instructions for use and storage. If making the milk plates in house, a couple of examples of how 3% (v/v) milk plates can be prepared are given here.

The first method for making up milk plates employs powdered milk and an agar base. Tryptic soy agar can be used, or you could use the agar on which the bacteria are being maintained. Once the agar is autoclaved, move the bottle containing the agar to a water-bath set at 45–50°C. Prepare the milk powder while the agar is cooling. To do this, weigh out the milk powder. Enough powder will need to be weighed out so that when it is dissolved in sterile water it will give a double-strength solution. For example, if making 100 mL milk solution (to give a final milk plate at 3%), weigh out 6 g of milk powder and add 100 mL sterile water (this gives you a 6% milk solution). This solution then needs to be boiled to sterilise the milk powder. This can be achieved by either boiling over a Bunsen burner or heating the milk solution on a hotplate at 100°C until the milk has come to the boiling point. Be careful not to overboil the milk, because it can rapidly boil out over the edges of its container . Once the milk has been sterilised and the agar base has cooled, they should be added together in a 50:50 ratio. Continuing with the example, this would involve adding 100 mL of the 6% milk solution to 100 mL of molten agar base (this will give you 200 mL of 3% milk agar). The milk solution can be added to the molten agar using a serological or adjustable pipette. Once the milk has been added, swirl the mixture gently to ensure the contents are thoroughly mixed.

This type of plate can also be made by adding 3% (w/v) milk powder to the agar before it is autoclaved (use single-strength agar for this; e.g. 3 g in 100 mL agar). However, in some cases, autoclaving the milk powder in the agar can give an uneven colouring to the agar plate when it is set, which makes it harder to interpret the protease activity of any bacteria that are plated on it.

These milk agar plates can also be made using sterile liquid milk instead of powdered milk, which can be added to autoclaved single-strength agar that has cooled to 50°C. When using sterile liquid milk, ensure the milk has warmed to room temperature before you add it, to avoid making the agar set before you have poured it. When the agar is cool enough, add 3% (v/v) sterile liquid milk (e.g. 3 mL milk to 97 mL molten agar) using either a serological pipette or an adjustable pipette. Flame the neck of the bottle containing the cooled molten agar and add milk to the agar. Flame the neck of the bottle again and replace the lid. Make sure the work is carried out aseptically and swirl the bottle to mix the agar and milk together thoroughly. Do not shake the bottle because it will create bubbles that will sit on top of the agar plate when you pour it. To pour the plates, stack the labelled Petri dishes close to the Bunsen burner. Starting at the bottom of the stack, pour 20 mL of the milk agar into each Petri dish. Leave the plates to set on the bench.

To test for casein activity, take a single colony of bacteria from an overnight agar plate or liquid culture and inoculate it onto the milk plate as a short streak or stab or in a well, as described for the blood plates in Section 8.1. Incubate the plates overnight at about 37°C, inverted if using a stab or streak and not inverted if using the well-based method. After incubation, inspect for activity. There will be one of two reactions after incubation:

1. **no protease activity:** The whole plate will be white and there will be no zone of clearing around the bacterial growth.
2. **protease activity:** There will be a zone of clearing in the plate around the bacteria growth.

If unsure about the activity, it can help to hold the plates up to the light or place them on a light box. As with the haemolysin tests, if unsure of what a positive and negative should look like, run a control un-inoculated plate alongside bacteria with known activity profiles to provide a comparison for positive and negative reactions.

Lipase and lecithinase activity - Lipase and lecithinase enzymes are secreted by bacteria and are capable of breaking down lipids within the environment. The lipase enzyme is capable of breaking down lipids and provides the bacteria with glycerol and smaller fatty acids that can be used by bacteria for a variety of processes such as energy production. The lecithinase enzyme breaks down lecithin and is thought to have a role in degrading host tissue during infection. There are ways to test for lipase activity including fluorescence assays (Lomolino et al., 2012; Unno et al., 2017); this section provides an example of a basic agar plate assay that can be carried out both rapidly and cheaply.

A test for both lipase and lecithinase activity can be carried simultaneously on an agar plate containing a source of fats for the bacteria to break down. Egg yolk is often used as a fat source that can easily be incorporated into an agar plate creating egg yolk agar. Egg yolk agar can either be purchased from a supplier such as Hardy Diagnostics (http://hardydiagnostics.com/) or it can be made using liquid egg yolk, which is then incorporated into an agar base. Liquid egg can also be purchased from various suppliers such as Oxoid (www.oxoid.com). A brief method outlining how to make this agar and test for lipase and lecithinase activity is given next. To perform this test, you will need bacterial isolates, agar, inoculating loops, a pipette, a Bunsen burner, a water-bath, a waste pot, sterile Petri dishes and liquid egg.

To make egg yolk agar plates, autoclave the agar base you intend to use, and then cool it to 50°C in a water-bath. Warm the liquid egg to room temperature while the agar is cooling. This prevents the agar from setting before you are ready. When the agar is cool enough, add 5% (v/v) sterile liquid egg (e.g. 5 mL egg to 95 mL molten agar) using either a serological pipette or an adjustable pipette, as outlined for the milk plate method. Make sure the work is carried out aseptically. Once the egg has been added to the agar, gently swirl the bottle to mix the agar and egg together thoroughly. Do not shake the bottle because it will create bubbles that will sit on top of the agar plate when you pour it. To pour the plates, stack the labelled Petri dishes close to the Bunsen burner. Working aseptically, pour 20 mL of the egg agar into each Petri dish. Leave the plates to set on the bench. Once set and dry, these plates can be stored in the refrigerator for 1–2 weeks.

To test for the lipase activity of a test bacterium, take a single colony of bacteria from an overnight agar plate or liquid culture and inoculate the egg plate as a short streak, stab or well, as described for the blood plates in Section 8.1. Incubate the plate overnight, inverted if using a stab or streak but not inverted if using the well method, at about 37°C before inspecting it for activity. There will be one of three reactions after incubation:

1. **no lipase/lecithinase activity:** The whole plate will be white and there will be no zone of clearing around the bacterial growth.
2. **lipase activity:** There will be a zone of iridescence in the plate around the area of bacteria growth. It can help to tilt the plate in the light to observe this effect. It will have an oil-like sheen.
3. **lecithinase activity:** There will be an opaque white or cream zone around the bacteria (Fig. 8.6).

Tributyrin agar is also commonly used to detect lipase activity from bacteria. This type of agar is available from various manufacturers and should be prepared according to their instructions and inoculated with bacteria as described in this section for the egg yolk plates. In terms of results after incubation, there should be:

1. **no lipase activity:** An opaque (cloudy white) appearance to the whole plate; or
2. **lipase activity:** A zone of clearing around the bacterial growth (Fig. 8.7).

Ensure you are clear about which media you are using, because the results differ depending on which media are employed.

It is particularly important when running these tests to run an un-inoculated egg yolk plate, because lecithinase activity can be hard to see; it can diffuse a long way through the agar. Having an un-inoculated plate in this

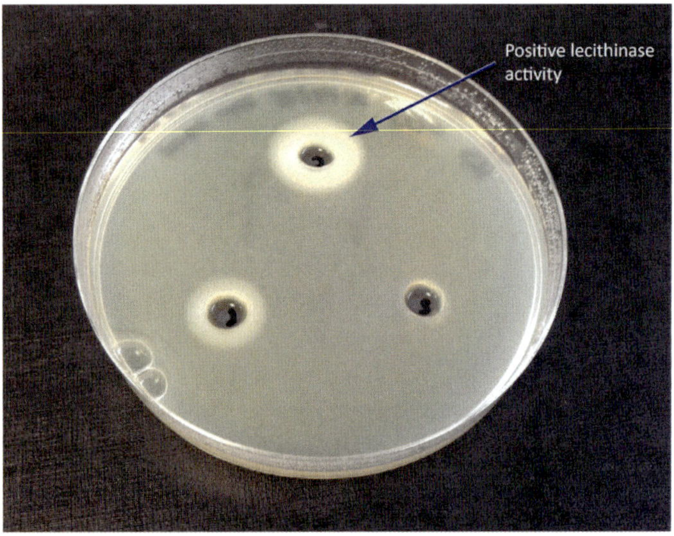

Figure 8.6 Egg plate after 24 h incubation at 37°C with bacteria inoculated into wells in the plate face demonstrating strong lecithinase activity.

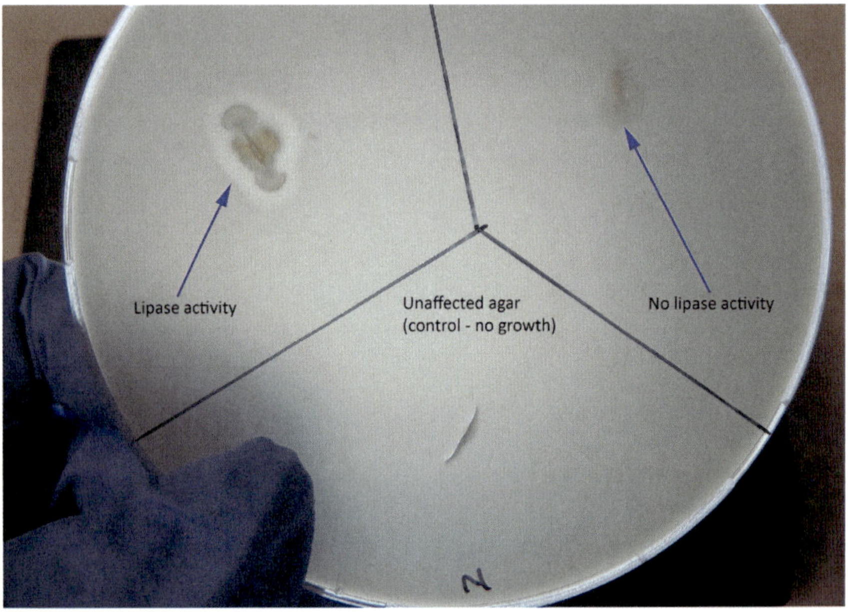

Figure 8.7 Tributyrin agar plate after 24 h incubation at 37°C with bacteria inoculated by stab into the agar demonstrating lipase activity.

case makes it easier to identify a diffuse reaction. It is also useful to check the manufacturer's website when using premade plates, because often there will be a list of control organisms that will give positive or negative reactions for each of the tests described. Using the control organisms alongside a test bacterium can help with correctly identifying the result and confirming that the media are working correctly.

DNase activity - The test outlined here will allow you to test the ability of an organism to produce deoxyribonuclease (DNase). This is an enzyme that breaks down DNA into smaller parts. DNase has been linked to the ability of bacteria to break down extracellular DNA and spread through host tissue, so it could be useful to measure DNase activity when looking at virulence in bacteria. The measurement of DNase activity is another test that can be done using an agar plate method. Pre-prepared DNase agar can be purchased from suppliers such as Oxoid (www.oxoid.com) and VWR (www.vwr.com). This agar can come as premade plates in Petri dishes with or without an indicator (such as methylene green) or as a powdered DNase agar base with an indicator added. If making the agar yourself, follow the manufacturer's instructions. As well as the agar, you will need bacterial isolates, inoculating loops, a pipette, a Bunsen burner, a waste pot and sterile Petri dishes.

If you are using DNase agar without an indicator, hydrochloric acid (HCl) will also be needed (1N HCl). Ensure appropriate personal protective equipment such as gloves and goggle are worn when handling HCl. Check your laboratory's Control of Substances Hazardous to Health Regulations 2002 risk assessment to see what safety measures it requires you to follow when handling this liquid. To test for DNase activity, take a single colony of the bacteria from an overnight agar plate or liquid culture and inoculate it onto the DNase plate as a short streak, stab or well, as described for the blood plates in Section 8.1. Incubate the inverted plate overnight at about 37°C. If using DNase agar with no indicator, the plate will need to be flooded with 1N HCl. This can be done using a large-volume pipette. Ensure the surface of the agar is covered and leave it to stand on a benchtop for a few minutes. Do not leave it for longer than 5–10 min. Once the plate has stood for a few minutes ,the excess HCl can be removed into a waste bottle before the plate is inspected. If the HCl is added too rapidly to the plate, it can wash the bacteria off the plate, which can make it harder to identify the activity, so try to add and remove the HCl gently. If leaving the HCl on the plate while the inspection take place, be careful not to tip the plate while inspecting it, because you do not want to spill the HCl. Inspect the plate. It can help to hold the plate against a dark surface. One of two reactions will occur after incubation:

1. **no DNase activity:** The whole plate will be a uniform colour and there will be no zone of clearing around the bacterial growth. The opaque appearance of the plate is caused by precipitation of the DNA within the agar on addition of the HCl.
2. **DNase activity:** There will be a zone of clearing in the plate around the bacterial growth (Fig. 8.8).

If DNase plates with an indicator have been used, there is no need to use the HCl. DNase plates with an indicator can be removed from the incubator and inspected directly. Check the manufacturer's instructions if you are using media with an indicator already added, to see what colour changes are expected. For example, when using methyl green, two reactions may could be observed:

1. **no DNase activity:** The whole plate will be a uniform colour (green in the case of methylene green) and there will be no zone of clearing around the bacterial growth

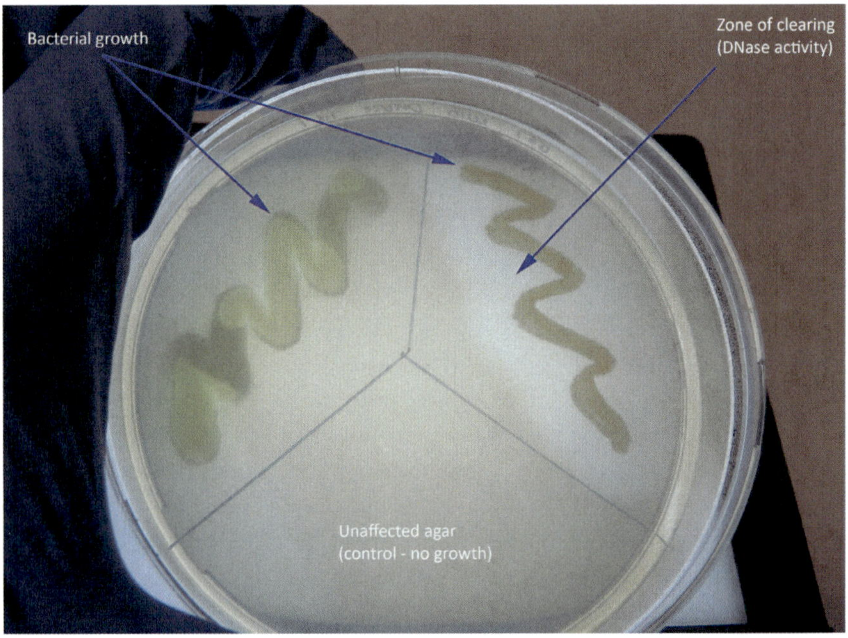

Figure 8.8 Deoxyribonuclease (DNase) agar plate after 24h incubation at 37°C with bacteria demonstrating clear zones (DNase activity) of clearing around bacterial growth.

2. **DNase activity:** There will be a zone of clearing in the plate where there is no green around the bacterial growth.

Adding HCl to the plate can kill some bacteria (such as Staphylococci), so when using this method, it would not be possible to re-incubate the plate after the initial inspection. If unsure about the activity of the organism, either prepare replicate plates that can be inspected at different time points or check that the bacteria show heavy growth before adding the HCl.

When using agar plate-based methods such as the ones described in this section, it is also possible to perform a quick screen of the activity of the bacteria being tested by adding a drop of liquid bacterial culture to the surface of the agar plate and incubating it overnight. When using this method, ensure that the drop of bacteria has dried into the plate before incubating; otherwise, when inverting the plate, the drops can run across the plate, making it difficult to interpret results. This is a useful technique for screening activity because it allows detection of a positive or negative reaction. However, it is difficult to quantify more subtle changes in activity using this method.

8.1.2 Non-agar plate-based assays

Catalase – As the name suggests, the catalase test allows you to determine whether an organism can produce catalase. This is an important virulence factor for microorganisms, because catalase can break down hydrogen peroxide into oxygen and water. The ability to do this prevents the death of bacteria through an accumulation of hydrogen peroxide created by some bacteria during aerobic respiration and reduces damage to bacteria from reactive oxygen species, such as those directed against bacteria by the host immune response. The test to see whether bacteria can produce catalase is straightforward and can be carried out using either a glass microscope slide method or an agar-based method.

To carry these tests, out you will require a glass slide, a glass coverslip, 3% (v/v) hydrogen peroxide, gloves, an overnight culture of bacteria on an agar plate, a plastic inoculating loop, a Pasteur or manual pipette, a sharps bin, a waste pot and a Bunsen burner. If using the agar-based method, the glass coverslip and microscope slide will not be needed but an agar plate or an agar slope inside a test tube (not blood agar) will be necessary.

To test for catalase activity using the microscope slide method, pipette a drop of H_2O_2 onto a glass coverslip and then aseptically transfer one isolated colony of bacteria from the agar plate onto the microscope slide. Carefully invert the microscope slide and slowly lower the bacteria on the microscope slide into the H_2O_2 on the coverslip. Try to keep the bacteria in the middle of the drop of H_2O_2. Observe the reaction at the point where the bacteria and H_2O_2 come into contact.

If using the agar based method, streak the bacteria of interest onto an agar plate or the agar slope. One streak will suffice (you are not streaking for single colonies). Incubate the bacteria. Once it has grown, add H_2O_2 to the bacteria on the agar. Again, observe the reaction as the H_2O_2 and bacteria come into contact with one another. One of two reactions will occur when the bacteria come into contact with the H_2O_2:

1. **no catalase activity:** There will be no obvious reaction indicating that the isolate being tested cannot produce catalase.
2. **catalase activity:** Bubbles will immediately be produced, demonstrating that the isolate being tested has catalase activity. The bubbles can be produced quickly. They can be hard to spot using the slide and coverslip method, because if the coverslip and glass slide are put together too quickly, the bubbles may become squashed and hard to see.

It can help to carry out the experiment against a dark-coloured background, because the bubbles can be hard to see when they are generated against a light background.

The microscope slide method can be done using just the drop of H_2O_2 on a slide with the bacteria added directly to it using a wooden stick or plastic loop. Adding the coverslip is for safety to contain bubbles that are produced. However, adding the H_2O_2 onto a bacterial colony that had been transferred to a glass slide, and not using a coverslip, has a small risk that the bubbles of H_2O_2 could come up off the slide. Do not use a metal inoculating loop, because the metal can react with H_2O_2 and create a misleading result. Make sure the bacteria are not taken from a blood agar plate, because there could be catalase activity from the blood cells in the agar, which would produce a false positive if these come in contact with the H_2O_2.

Coagulase activity – The enzyme coagulase allows bacteria to convert fibrinogen in host plasma to fibrin. This allows bacterial aggregation, which in turn is more difficult for the host immune cells to phagocytose and therefore increases the bacterial survival rate, potentially increasing the chance of infection. The test for coagulase is most commonly carried out to enable the differentiation of coagulase-positive *Staphylococcus aureus* and coagulase-negative *Staphylococcus epidermidis*. Various methods are available to test for coagulase activity, including using plasma, in which bacteria that produce the coagulase enzyme will cause fibrinogen to convert into fibrin appearing as clumps in the solution. Plasma can be sourced from companies such as Sigma-Aldrich (https://www.sigmaaldrich.com/united-kingdom.html) and can be bought as a solution, in tubes or on slides. However, there are also kits that will enable detection of the coagulase enzyme Protein A and capsular polysaccharide (found in methicillin-resistant *S. aureus*). The use of one of these tests, the Staphytect Plus test, will be described here.

To carry out the Staphytect Plus test you will need an overnight bacterial culture, the Staphytect Plus kit containing test latex, control latex and reaction cards (Fig. 8.6), a Bunsen burner, inoculating loops/sampling sticks and a waste pot.

To test for coagulase activity, bring the test reagents to room temperature and add a drop of test latex to the centre of one of the red circles on a test card. Repeat this on a second test card using the control latex. Using a sterile inoculating loop or sampling stick, pick a colony from the agar plate, mix this into the test latex and then repeat it for the control latex. Pick up the test card and gently roll the latex around within the circles. A rapid reaction should occur within around 20 s and give one of two results:

1. **no coagulase activity:** No obvious reaction liquid will remain; it will be smooth with no lumps.
2. **coagulase activity:** There will be rapid agglutination, which will appear as clumps within the liquid, indicating that the bacteria tested have coagulase activity, Protein A or capsular polysaccharide (Fig. 8.9).

To check whether the kit is functioning correctly, ensure that controls are used, and to check what positive and negative reactions look like, a known *S. aureus* can be run to provide a positive response, as well as a

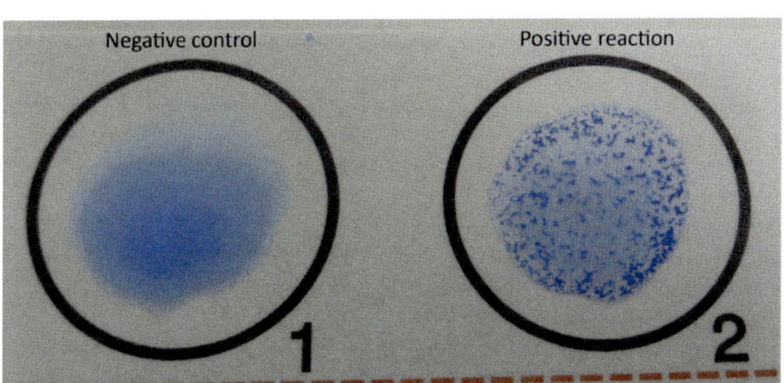

Figure 8.9 Negative (left) and positive coagulase tests (right). Note the clumping that occurs during a positive reaction.

known *S. epidermidis* as a negative control. Sometimes if a large number of bacteria are emulsified into the test reagent, they can break up into lumps within the latex. Be careful not to confuse this with a positive reaction. If in doubt, run the test a second time.

Oxidase activity - The oxidase test allow the differentiation of bacteria that produce cytochrome c oxidase and those that do not. The ability to produce the cytochrome c oxidase enzyme normally means that bacteria are able to use oxygen as a terminal electron acceptor for respiration. A number of tests can be performed to determine the oxidase status of bacteria. The filter paper method will be outlined here because it is one of the quickest and easiest.

To carry out the oxidase test, you will need an overnight bacterial culture on an agar plate, filter paper, an oxidase reagent (or filter paper containing reagent), gloves, a plastic inoculating loop, a sharps bin, a waste pot and a Bunsen burner. The oxidase reagent can be purchased from suppliers such as BD Diagnostic Systems (http://www.bd.com/europe/) or Sigma-Aldrich (https://www.sigmaaldrich.com/united-kingdom.html). This reagent will identify bacteria that have cytochrome c oxidase enzyme activity. In the oxidase test reagent, a chemical such as *tetra*-methyl-*p*-phenylenediamine acts as an artificial electron donor for cytochrome c; upon oxidation, it will change colour (a positive reaction).

To test for oxidase activity using the filter paper method, you will need to undertake the following procedure. When using an oxidase reagent that comes in a reagent dropper, you will first need to break the glass vial within the dropper. To do this, hold the dropper upright and squeeze gently until you hear the glass break. When inverted and gently squeezed, the reagent should come out drop by drop. Do not squeeze hard or it could cause the broken glass inside the dropper to pierce the bottle.

Drop the oxidase reagent onto filter paper and aseptically transfer an isolated colony of bacteria from the agar plate onto the filter paper, mixing it into the reagent with a sterile plastic loop or wooden stick. Observe the reagent for a colour change. The reaction should occur within about 30 s. Any reactions after this time should be discounted. If in doubt, repeat the experiments. The results should be one of two reactions:

1. **no oxidase activity:** No colour change or production of a light pink colour indicates that the isolate being tested cannot produce oxidase.
2. **oxidase activity:** A rapid change of colour to purple indicates that the bacterial isolate being tested can produce oxidase (Fig. 8.10).

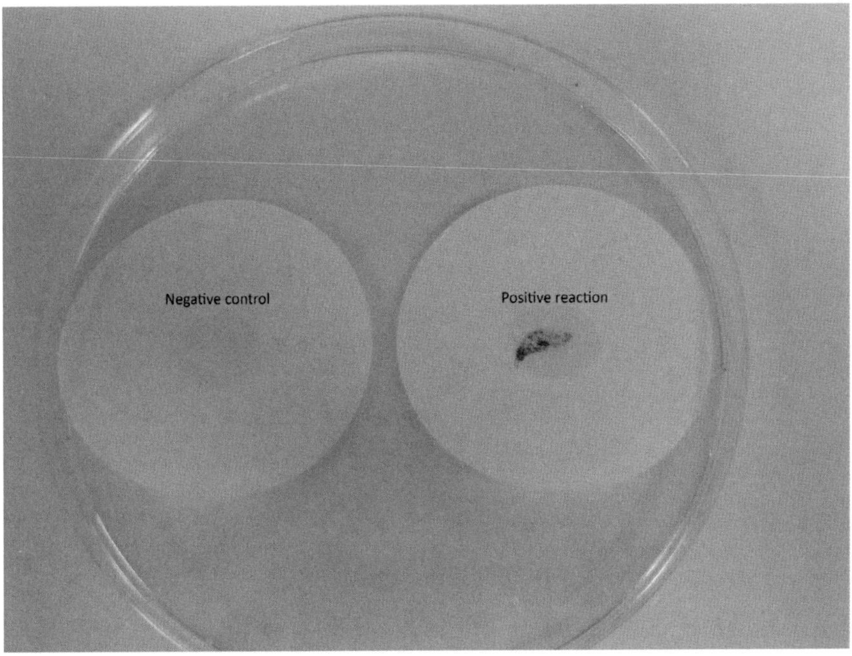

Figure 8.10 Example of a negative (left) and positive (right) oxidase test Pictures were taken within 30 s of the bacteria being inoculated.

It can help to run a bacterium with a known positive oxidase profile alongside the test bacteria if you are performing this test for the first time or unsure of what the positive reaction looks like. Having positive and negative control bacteria can also identify whether there are issues with the experiment.

Motility Test – Motility in bacteria is important because it allows migration of the organism to new sites within their environment (including the human host). They can move in response to various stimuli such as oxygen, nutrient or temperature gradients, which enables them to reach optimal sites for growth and division. The movement of many bacteria is often driven by flagella, which can be singular or multiple and located in various positions. Some bacteria can also move without flagella by gliding and twitching motility. Bacterial motility can be observed using microscopy techniques, which are outlined in Chapter 3. It is also possible to test for motility without a microscope. A common method is to test for motility in a test tube or on an agar plate using semi-set agar and is described subsequently.

To carry out a motility test, you will need an overnight bacterial sample, a test tube or agar plates with semi-solid agar, gloves, a stab inoculating loop and a Bunsen burner.

To test for motility, create semi-set agar in a test tube or agar plate. Motility test medium can be purchased from suppliers such as Sigma-Aldrich https://www.sigmaaldrich.com/united-kingdom.html or Mast Group Ltd. (http://www.mastgrp.com/); prepare it according to manufacturer's instructions. The media should be inoculated with the bacteria of interest by taking a colony of bacteria on a stab loop and stabbing the media. When using a test tube, ensure that the stab is in the centre and that it goes far enough into the media that the bacterial growth will be easily observable. When using agar plates, prepare them aseptically and then inoculate them by stabbing the bacterial culture into the middle of the agar plate. Incubate the bacteria at 35–37°C and inspect them after 24 h. Cultures can be re-incubated and inspected again after 48 h. When using the agar plate method, do not invert the plate for incubation or all of the agar will fall out of the plate. After incubation, you should be able to observe one of two outcomes:

1. **nonmotile:** The bacteria will have grown along the line of the stab, indicating that they are not motile.
2. **motile:** The bacteria will have grown out away from the stab line, indicating that they are motile. They will appear as diffuse growth through the tube, radiating away from the original site of inoculation (Fig. 8.11).

Media supplied from a manufacturer normally provide the identity of positive (motile) and negative (nonmotile) control bacteria to run alongside

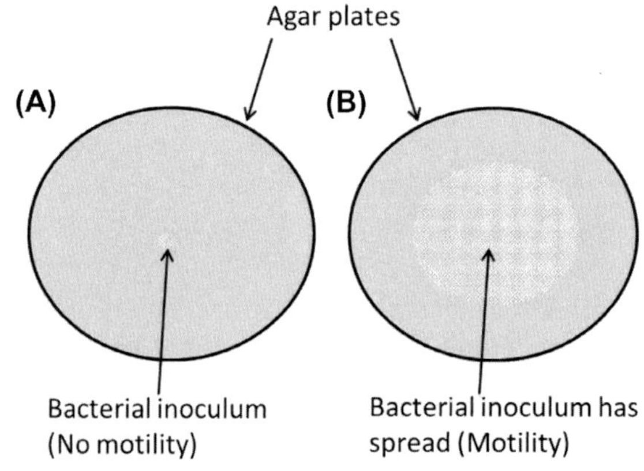

Figure 8.11 Diagrammatic representation of what might be seen when observing (A) nonmotile bacteria on an agar plate after 24 h incubation and (B) bacterial motility on an agar plate after 24 h incubation.

test samples. Suggestions for a positive control organism that has flagellated movement and should move away from the stab line might be an *Escherichia coli* and for the negative control something like an *S. epidermidis*; however, these are just suggestions and it is best to check the manufacturer's guidelines.

Some available motility media have dye incorporated into them that changes from colourless to coloured when taken up by viable bacteria. This can make it easier to identify whether the bacteria have grown away from the initial inoculation point. When employing the agar plate method, you can use callipers to measure the extent of the motility and the motility agar could be used to test the impact of various compounds on bacterial motility: for example, by comparing the extent of motility of bacteria after exposure to novel compounds or knocking out genes.

8.1.3 Virulence genes

When investigating the virulence potential of a bacterium, it is likely that you will want to look at the genes that the bacteria holds and potentially the changes in expression of those genes in response to different situations. For instance, the ability of a bacterium to express a toxin or quorum-sensing molecules may be particularly interesting. Or it could be that testing the effect of a new compound on the expression of a range of virulence gene expression is pertinent to your work. In these cases, molecular techniques could be used to investigate these questions. This will be covered in more detail in Chapter 9.

When considering the virulence potential of a bacterium, it is important to think about which virulence factors are normally seen in the bacteria you are studying and how they might change under different conditions. It is also useful to think about whether you would expect to see an increase or decrease in those factors under the conditions you are testing. The techniques provided in this chapter should allow you to start testing some of the most common virulence factors observed in bacteria and to optimise those methods for your own needs.

8.2 Notes page

Record observations and notes here.

CHAPTER NINE

Community composition studies

Contents

9.1 Introduction

In the laboratory, bacteria are usually cultured on agar to give pure colonies of a single species, and most microbiological studies have relied on this approach. However, in nature, bacteria thrive as polymicrobial communities in the environment and within their hosts. The complex interactions among different members of the microbial community can have an impact on survival, the capacity to colonise, virulence and the growth rate. Elements of the environment can also influence microbial community composition. This is rarely reflected by growth in complex or chemically defined media in a laboratory. Methods for studying microbial communities exist and try to take some of these factors into account. However, it is difficult to reproduce exactly within a laboratory setting the environment in which bacteria naturally grow. This is especially difficult if the environment is a living host such as a human or other mammal. Various techniques are available that rely on a combination of growth, molecular analysis and microscopy to understand microbial communities; some of these are described in this chapter.

9.1.1 Using culture methods to study microbial communities

A simple way to begin to dissect the composition of a microbial community is to use the total viable count (TVC) to indicate the different types and number of bacteria in a given sample. Using a combination of nonselective and selective or differential media, it is possible to isolate a variety of different bacteria

Bacteriology Methods for the Study of Infectious Diseases
ISBN 978-0-12-815222-5
https://doi.org/10.1016/B978-0-12-815222-5.00009-2

from one environmental sample, whether it is a soil sample or a wound swab. To do this, it is important first to use nonselective media to enrich all of the bacteria in the sample and to prepare replicate plates so that one can be cultured aerobically and the other anaerobically. Using this approach, it will be possible to recover a variety of bacteria that can be individually identified by subculturing onto selective or differential media with subsequent biochemical testing. However, a well-acknowledged drawback to this approach is that not all members of a microbial community are recovered, and so the data derived are not a true indication of the type or number of microorganisms in the community. It might serve as a good starting point for community composition studies that can be further investigated with molecular techniques or by microscopy, so it should not be completely discounted in the first instance.

If you already know the types of bacteria typically found in the environment you are studying, it is possible to understand how members of the community interact using artificial community models. These have obvious disadvantages in that you are choosing to study only certain bacteria under specific laboratory conditions, but they can provide to preliminary data with which you can begin to understand more complex interactions. How to use artificial community models and the type of data that can be gathered from them will be outlined in this section.

A simple starting point for analysing the types of bacterial interaction that could occur among known members of a microbial community is to assess paired interactions. To do this, first compile a list of the bacteria you expect to find in the environment in which you are interested: for example, all of the bacteria known to colonise an infected lung in someone with cystic fibrosis. Next, allocate the bacteria into pairs, making sure that every different combination of pairs of bacteria is accounted for. Culture each bacteria in broth until they are confluent, and then pellet them by centrifugation. Wash the pellet twice in coaggregation buffer (1 mM Tris [pH 8], 150 mM NaCl, 0.1 mM $CaCl_2{\cdot}2H_2O$, 0.1 mM $MgCl_2{\cdot}6H_2O$) by resuspending and centrifuging (Cisar et al., 1979). The next step is to equilibrate each of the two bacteria you intend to study so that they are equal to each other. To do this resuspend each pellet in the same volume of coaggregation buffer and read the optical density (OD) in a spectrophotometer at 600–650 nm. Slowly keep adding buffer until the OD is the same for each bacterium (aim for an OD for each bacterium of 0.09–1.0). Place the equilibrated cells on ice. When you are ready to begin the assay, add 0.5 mL of each bacterial suspension to a semi-microcuvette and immediately take a reading at 600–650 nm. Incubate the cuvette at room temperature and take additional readings at 20–30-min

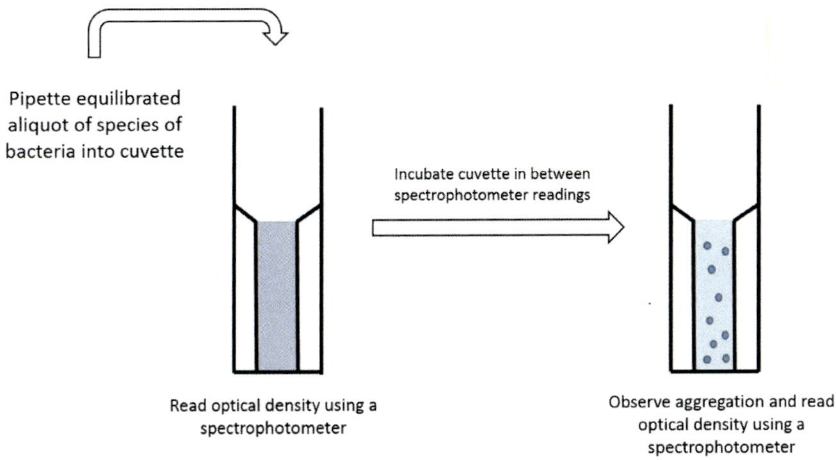

Figure 9.1 Schematic representation of coaggregation assay. Clumps of co-aggregating bacteria form over time and can be observed by eye and measured by spectrophotometry.

intervals. If the two bacteria interact, you should see by eye that they begin to clump together (Fig. 9.1). By taking a reading with a spectrophotometer, you can quantify how much the bacteria aggregate over time and calculate this as a percentage of the reading taken at time 0. This will enable you to see which pairs of bacteria interact and which seem to interact better, or at least coaggregate more quickly or to a greater or lesser extent. If you know something about the environmental conditions of the bacteria you are studying, this experiment could be done using, for example, artificial sputum, simulated wound fluid, artificial saliva, etc., because the different components of such media might affect the capacity of bacteria to interact and reflect better how they behave in their usual environment.

Still using pairs of bacteria, an alternative approach might be to assess how well they grow together. Consider the techniques you have encountered in Chapter 2 and Chapter 6 to decide the best approach. For example, do the bacteria you are studying ordinarily grow in suspension or as a biofilm? If you decide on a broth culture approach, it will be necessary to grow and equilibrate precultures as described for the coaggregation experiment (but resuspended in media) and to inoculate the growth media with equal numbers of the two bacteria. You could try a series of different ratios to assess the impact this has on the composition of the two-species community. To determine how the two bacteria behave in co-culture, it will be necessary to take samples at given time points (e.g. 24, 48, 92 h) and analyse them by TVC (Fig. 9.2). You will most likely need to use two different types of

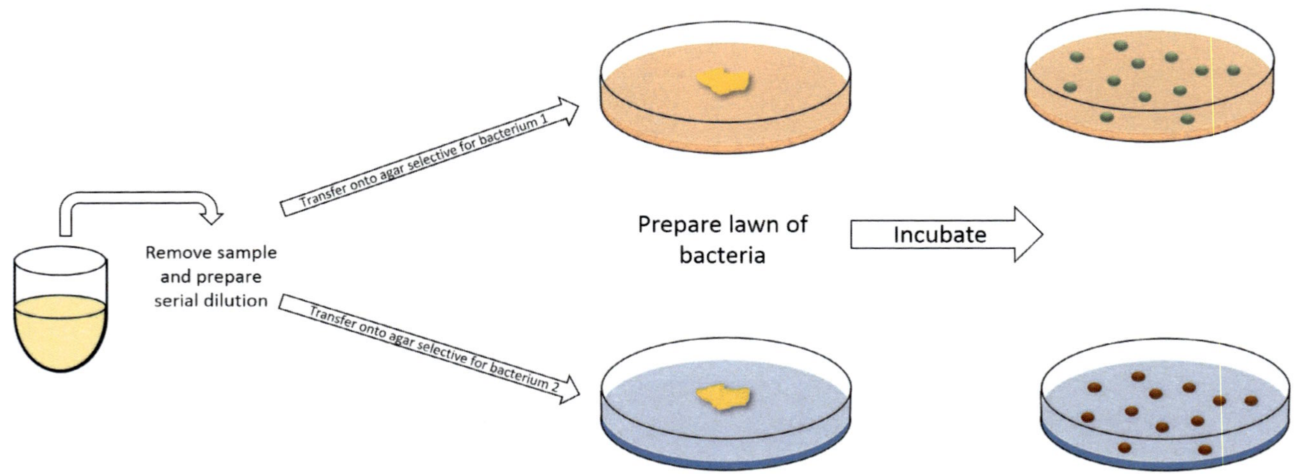

Figure 9.2 Enumerate bacteria from mixed culture by plating onto two different selective agars that allow only for the growth of each single species.

media to recover and differentiate between the different bacteria. Once you have data for two species, it will be possible to add more, based on the results gained from the coaggregation experiments. An important point to remember is that this type of experiment uses a batch culture method in which nutrients are depleted over time, so be mindful of this when choosing the time points for analysis and interpreting the data if you see the microbial composition changing. Once you have the preliminary data from this type of experiment, you could consider using a flow system to avoid effects cause by nutrient depletion. Other factors to consider might be the composition of the media; these can be changed to take account of pH, micronutrients and salts that would be ordinarily found in the natural environment of the bacteria being studied. Alternatively, artificial fluids could be used, such as those mentioned for the coaggregation assay.

If you are taking a biofilm model approach, the setup of the experiment will be similar to that described earlier with regard to equilibrating the inoculum, inoculation rations and time point sampling. Again, consider whether it would be best to use a batch or flow system and the type of media to use. For both types of experiment, it will be easiest to begin with a two-species system and then progress to multi-species experiments.

The TVC data derived from these experiments (see Chapter 6) will tell you how many of each bacterium were recovered at each time point. By equilibrating the original inoculum, you will know how many bacteria you added into the experiment and at what ratio (Table 9.1).

This will allow you to determine how the ratio of bacteria changed over time and which bacterium grew best, based on the numbers recovered. If you

Table 9.1 Optical densities and approximate cell number to which these equate when read in a spectrophotometer at A_{600}.

Optical density at A_{600}	Approximate number of cells (colony forming units/mL)
0.08–0.1	1.5×10^8
0.14–0.17	3.0×10^8
0.27–0.31	6.0×10^8
0.38–0.42	9.0×10^8
0.51–0.55	1.2×10^9
0.67–0.7	1.5×10^9
0.74–0.77	1.8×10^9
0.83–0.88	2.1×10^9
0.94–0.98	2.4×10^9

Values are based on the McFarland standard system.

have data for several different start ratios and different types of media, you will begin to develop a salient knowledge of the sorts of factors that can affect the composition of the community. It is possible to quantify and pictorially represent these changes by calculating the competitive relative index for each bacterium in the community. The competitive relative index calculates a number based on the ratio of the input and output populations relative to each other. The higher the number, the better the bacteria are at competing in a given environment. The threshold is taken to be 1; values above 1 show good competition, and below 1, poor competition. Table 9.2 demonstrates how to calculate the competitive relative index for a two-species experiment. Fig. 9.3 shows how these data can be presented and interpreted.

9.1.2 Using culture methods to assess rates of mutation within a microbial community

The impact of interspecies interaction on evolution can be ascertained using the same simple two-species models as described earlier. In this case, the focus will be on the effect of a multi-species community on antimicrobial resistance and can be tied into the experiments described in Chapter 4. This type of analysis uses the Luria-Delbr*ü*ck fluctuation test based on the bacterial phenotype rather than genotypic evaluation. Therefore, the mutation rate is inferred from the phenotype, but it could be verified by genome analysis. An example of how to set up an experiment and analyse the data using the Luria- Delbr*ü*ck fluctuation test is described next.

You will need to prepare single-species cultures for each bacterium and grow them in media for 16–24 h. These must then be equilibrated to a known colony forming unit (CFU) (as described in Section 9.1) and the co-cultured bacteria mixed in a 1:1 ratio (as a start point, but a range can be studied). Prepare 12 replicates in 1–5 mL media for each single species and co-cultured pair and incubate them at 37°C for 24 h. Pellet the bacteria by centrifugation and prepare a serial dilution (as described in Chapter 2 and Chapter 5). Plate onto selective media to recover each specific species of bacteria and also plate replicates on the same selective media containing a specific antibiotic (identified in Chapter 4) at 2 × *minimum* inhibitory concentration (MIC) 2 and 4 × MIC (Fig. 9.4). Incubate the plates at 37°C; enumerate the antibiotic-free plates after 24 h and the antibiotic-containing plates after 48 h. Prepare an Excel spreadsheet of the CFUs recovered and calculate an average for each. Use the following online calculator to determine the mutation rate based on the average CFU values: http://www.mitochondria.org/protocols/FALCOR.html (Hall et al., 2009).

Table 9.2 Example of how to calculate W (the competitive relative index) for Bacterium A and Bacterium B cultured together over a time frame of 96 h, with samples recovered and enumerated at 24-h intervals.

	Co-cultured bacteria							
	24 h		48 h		72 h		96 h	
	A	B	A	B	A	B	A	B
Rf	8.90E+19	2.87E+19	8.9E+19	2.87E+19	8.9E+19	2.87E+19	8.9E+19	2.87E+19
Sf	2.87E+19	8.90E+19	2.87E+19	8.9E+19	2.87E+19	8.9E+19	2.87E+19	8.9E+19
Ri	1.46E+22	2.65E+21	1.40E+23	1.81E+23	7.97E+24	8.30E+24	3.20E+25	2.70E+26
Si	2.65E+21	1.46E+22	1.81E+23	1.4E+23	8.3E+24	7.967E+24	2.70E+26	3.2E+25
Rf/Ri	6.09E–03	1.08E–02	6.36E–04	1.59E–04	1.12E–05	3.46E–06	2.78E–06	1.06E–07
Sf/Si	0.01083	0.006093	0.000158,564	0.000636	3.458E-06	1.117E-05	1.063E-07	2.78E-06
ln(Rf/Ri)	−5.10054	−4.52542	−7.360761332	−8.74936	−11.40214	−12.57487	−12.79261	−16.057
ln(Sf/Si)	−4.52542	−5.10054	−8.749355187	−7.36076	−12.57487	−11.40214	−16.05704	−12.7926
W	**1.127087**	**0.887243**	**0.841291864**	**1.188648**	**0.9067403**	**1.1028516**	**0.7966981**	**1.255181**

Rf and Sf are the start colony forming units (CFU) for the cultures. For example, in the first column, Rf is the start CFU for Bacterium A and Sf is the start CFU for Bacterium B. This is the opposite way around for the second column. Ri and Si are the CFU from the biofilm that was cultured. For example, in the first column, Ri is for Bacterium A and Si is for Bacterium B; it is the opposite way around for the second column.

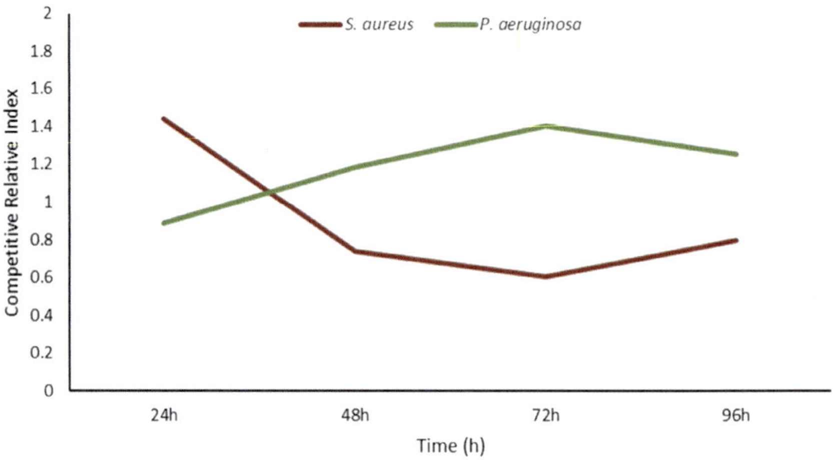

Figure 9.3 Example of graph showing the relative competitive index for a two-species culture of bacteria.

These experiments can also be done with biofilms cultured in 24-well microtitre plates (see Chapter 6), but for enumeration, the biofilms must be scraped before to preparing and plating serial dilutions (see Chapter 6).

9.1.3 Analysis of microbial communities using molecular techniques

The methodologies described in Sections 9.1 and 9.2 are most useful for studying artificial microbial communities. They are fairly labour-intensive and provide salient information about how bacteria might interact in co-culture. In essence, they provide a simple model to describe some phenomena that might occur in complex microbial communities. This means that they are still applicable for early-stage research that can be built upon to develop a broader understanding of the role of microbial communities, such as in the progression of infectious disease. This section describes some molecular methods that can be used to analyse microbial communities. These can be applied to artificial microbial communities but are most commonly used to study natural microbial communities such as swabs or biopsies taken from wounds, scrapings of tooth plaque, sputum from the lungs of people with cystic fibrosis, soil cores, waste water systems and estuarine habitats, to name a few.

The 16s ribosomal RNA (rRNA) gene is conserved in all bacteria and has been used as a tool for phylogenetic analysis of bacterial communities since the 1990s. It contains nine variable regions and

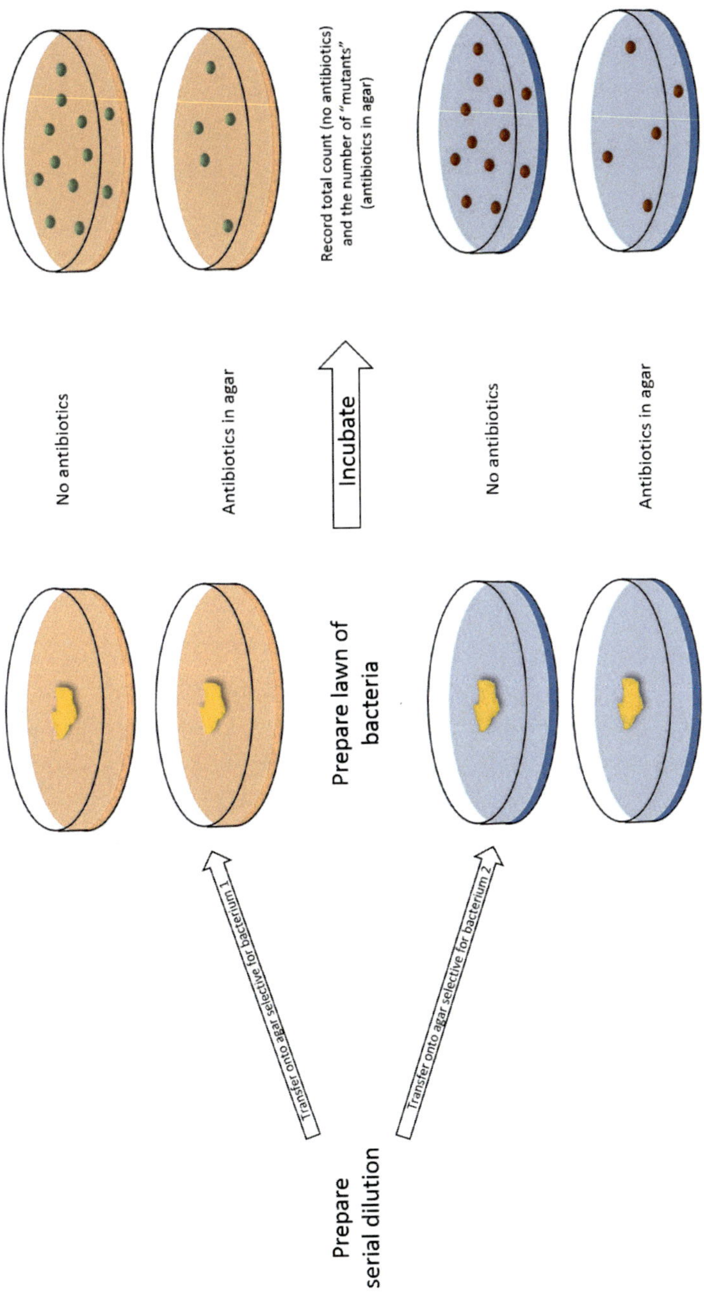

Figure 9.4 Schematic representation of Luria- Delbrück method for determining mutation rate within a microbial population.

Figure 9.5 Schematic representation of 16s ribosomal RNA gene. Variable regions are shown in *blue*.

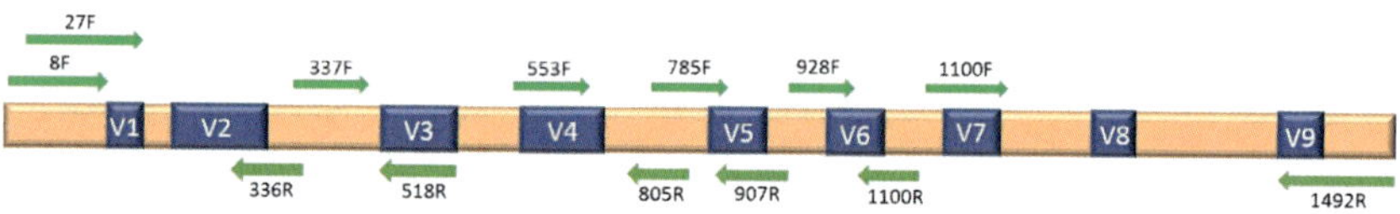

Figure 9.6 Schematic representation of 16s ribosomal RNA gene with universal primer sites indicated by *arrows* above and below the gene.

10 conserved regions (Fig. 9.5). 16s rRNA databases are publicly available (www.ncbi.nlm.nih.gov) and the sequences of the conserved regions of this gene are known.

Mixed environmental samples can be analysed by polymerase chain reaction (PCR) targeting the 16s RNA gene. Primers spanning different regions of the 16s rRNA gene can be used alone or combined with species-specific primers to detect members of a microbial community (Fig. 9.6).

Universal 16s rRNA primers can be used to identify broad groups of bacteria such as *Firmicutes* or *Proteobacteria*, and species-specific 16s rRNA primers can be used to identify known bacteria within the sample. The latter assumes knowledge of the members of the population, and therefore will identify and detect only known species, excluding unknown ones. If you are trying to capture all species in a given sample, a good approach might be to use universal primers and then analyse the PCR products by sequencing or denaturing gradient gel electrophoresis (DGGE). DGGE enables the separation of fragments of DNA based on a denaturing gradient of urea within the gel. Each fragment of DNA will denature at a different concentration of urea to produce a single strand of DNA anchored by a guanine-cytosine (GC) clamp. The specific denaturing concentration is determined by the nucleotide sequence of the DNA fragment, and so each fragment is separated on the gel according to its sequence rather than its size. This technique requires an additional round of PCR to attach the GC clamp to the PCR product. The GC clamps holds the two denatured strands of DNA together (Fig. 9.7).

A typical workflow for studying bacterial populations using the 16s rRNA gene is given in Fig. 9.8.

If you intend to use 16s rRNA to analyse a bacterial population, you will need to decide which approach is best, bearing in mind both

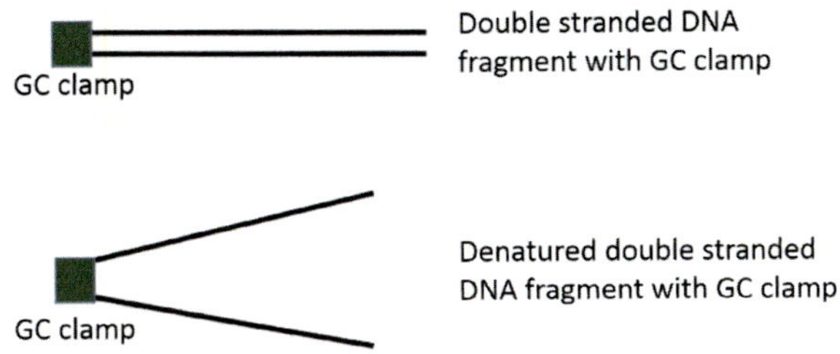

Figure 9.7 Schematic representation of a denatured DNA fragment with a guanine-cytosine (GC) clamp. The GC clamp prevents denatured DNA from completely separating into single strands.

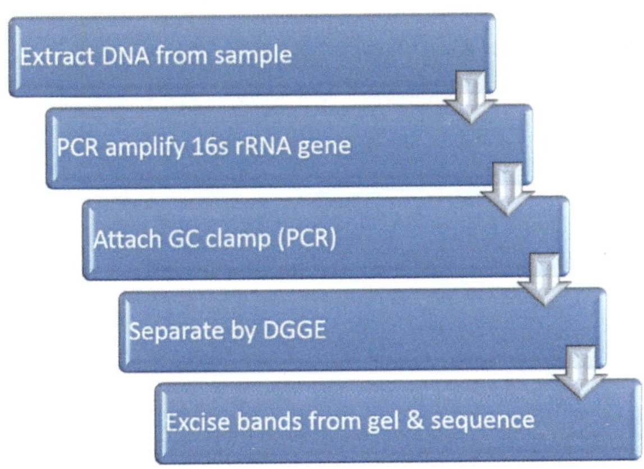

Figure 9.8 Flow diagram of typical workflow for denaturing gradient gel electrophoresis (DGGE) from DNA extraction to sequencing to identify species. *GC*, guanine-cytosine; *PCR*, polymerase chain reaction; *rRNA*, ribosomal RNA.

constraints of time and cost. A basic outline for DGGE using urea as a denaturing agent is given next as a guide for the approximate time frame of the experiment. The BioRad DGGE system is most commonly used and a detailed protocol to set up and run DGGE gels can be found at http://www.bio-rad.com/webroot/web/pdf/lsr/literature/M1709080C.pdf.

Day 1	First round of polymerase chain reaction (PCR) amplification of 16s ribosomal RNA gene using universal primers for one of the conserved regions of the gene Second round of PCR to attach the guanine-cytosine clamp (can be run overnight with a 4°C hold-step) Keep samples at −20°C until ready to analyse

Day 2	Cast gel; clean glass plates with detergent followed by distilled water and then ethanol. Dry using a lint-free cloth
	Assemble the plates: place spacers between the plates and affix gaskets and secure with casting clamps
	Ensure the gradient mixer is clean and check it is working by flowing water through it
	Prepare the gel mixes (the percentage of the gradient will depend on the experiment) and fill both chambers of the mixer (add the ammonium persulphate only when ready to fill the chambers)
	Turn off the pump when the gel is 3–4 cm from the top of the glass plates
	Overlay with distilled water and allow the gel to set (2–4 h). Pour off the water once it is set; add the stacking gel and comb
	Fill the gel tank with 1× *Tris-acetate-ethylenediaminetetraacetic acid* buffer and heat it to 60°C
	Transfer the gel to the gel tank frame and secure with clamps; insert the frame/gel into the gel tank
	Remove the comb and load the samples; run the gel at an appropriate voltage for 5–14 h (depending on the size of the gel)
Day 3	Remove the gel from the tank and stain it with ethidium bromide (Chapter 7), ensuring that personal protective equipment is worn throughout (ethidium bromide is a carcinogen and mutagen)

Quantitative reverse-transcriptase PCR is also powerful for analysing microbial populations. Details with regard to planning, optimising and analysing data generated from these experiments are discussed in Chapter 7 and will not be reiterated here. However, this section outlines some factors you will need to consider when applying this technique to the study of microbial populations. First, consider whether you want to use this technique to study an artificial microbial population or a naturally occurring one, such as a clinical or environmental sample. We will begin by considering an artificial population.

In this type of population, you will know the species of the bacteria, the ratio of the start inoculum and the type of culture model (these experiments will have been planned and optimised as described in Section 9.1). Therefore, you can decide at what time points you wish to analyse the population and which microorganisms you are looking for. An easy approach might be to begin with a two-species system and analyse it at 24-h periods over 72 h to match with the experiments devised in Section 9.1. For this approach, you can use species-specific PCR primers and a multiplex PCR assay (see Chapter 7). It is possible to assess populations cultured in any type of model provided you can recover the bacteria, and data derived in terms of microbial quantity and time-dependent shifts can be mapped against CFU data. DNA will be isolated from the entire population at three different time points, so it will be necessary to prepare three identical cultures, remembering that these have to be prepared with both biological and technical replicates.

At the appropriate time point, you will need to harvest the bacteria and extract the total DNA. You can use any of the numerous commercially available DNA extraction kits to do this or use your own protocol (see Chapter 7). It is a good idea to collect DNA from each time point and store it until you are ready to undertake PCR analysis. Because this is quantitative (Q)-PCR, it will be imperative to assess the DNA for both quantity and purity using a spectrophotometer, as described in Chapter 7. To ensure the data are comparative, equilibrate the DNA samples to the same concentration so that the data can be reliably compared and to ensure that any observed changes result from changes in the population and not from different concentrations of template DNA.

To assess the relative bacterial population over time, it is necessary to derive the cyclic threshold value from the relative fluorescence unit values produced from the PCR reaction for each experimental time point. It is possible to use comparative single-species studies and titrated DNA to determine the actual bacterial load, but if used in conjunction with CFU data, relative values will suffice as a start point. If you are planning to use Q-PCR instead of culture-based methods, determine the actual bacterial load. However, PCR will not discriminate between live and dead bacteria.

If you intend to use Q-PCR to analyse naturally occurring bacterial populations directly, such as clinical or environmental samples, you will need to use an approach that will allow you to capture all of the species, rather than the ones about which you know; i.e., if you only use species-specific primers, you will detect only those bacteria and miss others. Therefore, the use of broad-range 16S primers to detect groups of bacteria might be best instead of species-specific primers. To gather information regarding the identities of each species in the sample within these groups, sequence the PCR products and then conduct a further experiment with specific primers to assess individual members of the population. To detect rather than quantify the members of a given sample, end point PCR can be used before sequencing; the information gathered from these data can be used to design a panel of species-specific primers for more thorough quantitative analysis of the population. Such studies might exceed the scope of your research. Numerous published research articles are available that detail specific experimental design. Therefore, these will not be reviewed here, because they are outside of the remit of this book.

Table 9.3 Simulated body fluids for which recipes are available.

Simulated body fluid
Artificial blood
Artificial saliva
Artificial sputum
Artificial tears
Artificial urine
Simulated colonic fluid (fasting and fed)
Simulated gastric fluid
Simulated intestinal fluid (fasting and fed)
Simulated lung fluid
Simulated semen fluid
Simulated sweat
Simulated synovial fluid
Simulated vaginal fluid

9.1.4 Considering the impact of environment

When planning any of these experiments it is important to consider the normal environment in which the bacteria being studied live. For example, if they are clinical specimens, which site of the body were they originally from? Consider the temperature, pH, micronutrient availability, oxygen availability, nutritional sources, and so on that these bacteria might encounter, and then try to adapt your model accordingly: for instance, by using artificial body fluids. It might be interesting to optimise the model initially with a general bacterial growth medium and then assess what the population looks like under different environmental conditions that are appropriate to their normal environment. All of these different factors can be integrated into any of the models described from the outset. Table 9.3 lists some common artificial body fluid recipes that are available; the following reference has details of the ingredients and how to prepare these fluids (remember that they will need to be sterilised before use): Marques et al. (2011).

9.1.5 Summary

There are numerous ways to model and study microbial communities. It is important to consider which best suits the question you want to answer. Careful planning from the outset will allow you to include all of the appropriate variables and design experiments that enable you to gather the broadest and/or most applicable data to test your hypotheses. These models can be an extension of studies to assess novel antimicrobials, virulence and general growth characteristics and offer an alternative perspective to cell-culture based models.

9.2 Notes page

Record observations and notes here.

CHAPTER TEN

Invertebrate infection models

Contents

10.1 Introduction

In vitro infection models such as those described in Chapter 5 provide insight into how bacteria might interact with their host but are not truly representative of a whole living organism. Therefore, the conclusions drawn from such models are limited, with some room for extrapolation. Before moving from in vitro to in vivo models, it is important to consider the need to use living organisms to answer the research question. In vivo infection models can be complex and expensive, requiring specialised training, facilities and legislature. Invertebrate models offer an alternative approach that is more affordable and does not require specialist equipment, facilities or legislature. Drawbacks of using invertebrates are acknowledged, but they are widely used as a means of gathering valuable data regarding pathogenicity and virulence in a living organism, as well as screening for the efficacy of new antimicrobial interventions. This chapter describes two in vivo invertebrate infection models that provide a good first step toward facilitating the move from in vitro analyses.

Bacteriology Methods for the Study of Infectious Diseases
ISBN 978-0-12-815222-5
https://doi.org/10.1016/B978-0-12-815222-5.00010-9

10.1.1 Caenorhabditis elegans

Caenorhabditis elegans is a free-living nematode found in soil. It has an intestine and pharynx but lacks a respiratory and circulatory system. It is composed of a primitive, ancestral immune system responsive to challenge by viruses, bacteria and fungi and consequently has been used as a means to study the innate immune response to infection. As an infection model, *C. elegans* is further attractive because it is easy to grow in bulk and maintain, is inexpensive to use and can be stored as a frozen stock.

C. elegans can be purchased from the *Caenorhabditis* Genetic Centre (https://cbs.umn.edu/cgc/home) along with detailed protocols for a variety of experiments and applications. This chapter describes a protocol to assess the potential for a pathogen to kill *C. elegans* and how to test the effectiveness of an antimicrobial intervention.

Nematode growth media must be prepared for the growth and maintenance of *C. elegans* as follows:

Mix 3 g NaCl, 2.5 g peptone and 17 g of agar in 975 mL distilled water. Autoclave and allow to cool to 55°C. Add each of the following, and mix by gently swirling between adding each ingredient: 1 mL 5 mg ml^{-1} cholesterol (in ethanol), 1 mL 1M $CaCl_2$, 1 mL 1M $MgSO_4$, and 25 mL 1M potassium phosphate, pH 6.0 (Steirnagle et al., 1999). Once combined, pour into agar plates as per normal and allow them to set. Ensure that all of the additives are filter sterilised before use. You will need to use Petri dishes with a diameter of 3.5 cm.

C. elegans requires a food source; this is provided in the form of a bacterial lawn of *Escherichia coli* strain OP50. These are available for purchase freeze-dried from a variety of suppliers including Fisher Scientific and NemaMetrix. The strain will need to be recovered from the freeze-dried pellet according to the manufacturer's instructions, and a lawn plate prepared according to the method given in Chapter 2 and Chapter 4. Incubate the lawn overnight before adding the nematodes and transfer them to a fresh lawn plate each week to maintain them (Fig. 10.1).

Before undertaking the killing assays, it is necessary to synchronise the *C. elegans* culture so that they are all in the same growth phase (Fig. 10.1), which is thus: At 4 days before commencing the assay, as described by (Laws et al., 2004),

1. Wash the *C. elegans* culture plates by adding 5 mL distilled water; transfer the nematode suspension to a 50-mL Falcon tube.
2. Add 100 μL 5% sodium hypochlorite solution and 0.1 g sodium hydroxide.

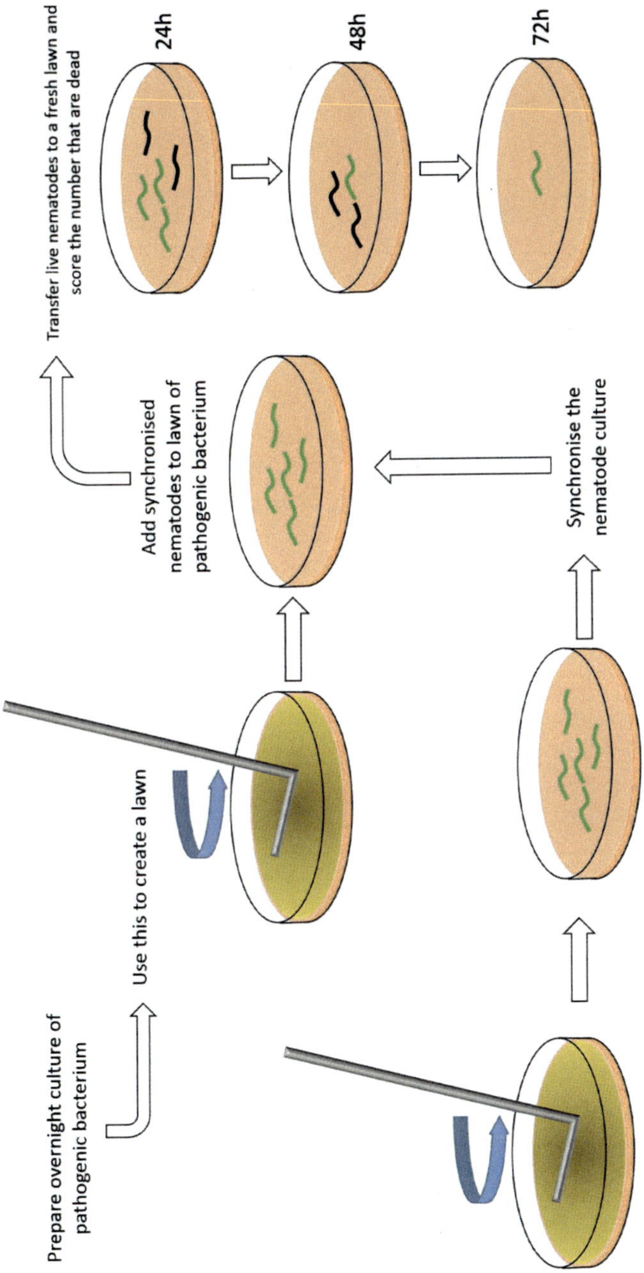

Figure 10.1 Schematic representation of process required to prepare infect and score *C. elegans* for survival to assess the pathogenicity of a bacterium.

3. Shake the suspension for 10 min at 100–200 revolutions per minute (rpm) at 25°C. This will kill all of the nematodes but not the eggs.
4. Add 45 mL distilled water and centrifuge for 15 min at 1000*g*
5. Carefully remove the supernatant and resuspend the pellet in 50 mL K medium (3.075 g NaCl, 2.42 g KCl in 1 L distilled water).
6. Incubate the suspension overnight at 25°C at 50–100 rpm. This allows the eggs to hatch.
7. Centrifuge the hatched nematodes (in the L1 stage of growth) for 15 min at 1000*g* and carefully discard the supernatant.
8. Inoculate fresh culture plates (of OP50) and incubate at 16°C for 3 days until the nematodes reach the L4 stage (this must be verified by light microscopy). They are then ready for use in the assay.

To carry out the killing assay, prepare the bacterial cultures you wish to investigate at the same time as you prepare the nematodes. To do this, prepare overnight cultures of the bacteria 2 days before the assay. At 1 day before the assay, equilibrate them to an appropriate optical density (OD) (Fig. 10.1). To determine the pathogenicity or virulence of bacteria with respect to the dose, it will be necessary to test a range of different dilutions or ODs, and therefore you will need enough plates of *C. elegans* for all of the different doses or dilutions.

Once the bacteria are equilibrated, pipette 20 μL into the centre of a 3.5-mL-diameter nematode growth media agar plate and spread to create a lawn. Incubate this overnight at 37°C. The next day, you will need to transfer 10–20 synchronous nematodes from the OP50 plate onto the lawn of bacteria you are testing. Do this using a worm-pick, which are available for purchase from www.wormstuff.com, and a dissecting microscope. This can be tricky to perform and may require practice. When the worms have been transferred, incubate the plates at 25°C. As described by Laws et al. (Laws et al., 2004) the nematodes will continue to reproduce at this temperature, so you will need to transfer the original and still living nematodes to fresh bacterial lawns every day of the experiment (begin with 24-, 48- and 72-h time points) (Fig. 10.1). Check each day whether the nematodes are alive or dead by touching them with a mounted needle or worm-picker. If the worms are not responsive to touch, they are scored as dead and need not be transferred (Fig. 10.1).

It is possible to use an adapted version of this assay to determine how readily *C. elegans* are colonised by bacteria and if they are able to clear the infection over time. At given time points (which can be determined from the killing assay), the nematodes can be harvested and the number of bacteria within the digestive tract determined by total viable count. The following method is based on that of Garsin et al. (Garsin et al., 2001) and Laws et al. (Laws et al., 2004):

1. At each time point, pick up five nematodes and transfer them to a brain-heart infusion agar plate containing 25 $\mu g\,mL^{-1}$ gentamycin.
2. Add 50 µL 25 mM sodium azide to the nematodes to remove surface bacteria and halt nematode excretion.
3. Place the nematodes in a sterile 1.5-mL microcentrifuge tube containing 20 µL M9 medium with 1% Triton X-100.
4. Mechanically disrupt the nematodes using a pestle.
5. Adjust the volume to 50 µL by adding M9/1% Triton X-100 and prepare a serial dilution for enumeration as described in Chapter 2 and Chapter 5.

10.1.2 Galleria mellonella

Commonly known as the wax moth, the larvae of *G. mellonella* are frequently used to study pathogenicity as an inexpensive and ethically unconstrained infection model. They can be easily and affordably purchased from pet shops (nonsterile larvae) or breeding laboratories (larvae bred under sterile conditions: for example, TruLarv: www.biosystemstechnology.com). If using the former, larvae can be surface sterilised by spraying with ethanol before the experiment take place. No special equipment is needed to handle and maintain the larvae and their short life span makes them ideal for screening bacteria and also antimicrobial treatments, the protocols for which will be described in this section.

G. mellonella does not have an adaptive immune system, but the innate immune system is similar to that of vertebrate organisms. For infection experiments, suspensions of bacteria are injected, via the proleg (Fig. 10.2), into the haemocoel, which is filled with haemolymph that contains the constituents of the immune system.

Much like *C. elegans*, this model can be used to determine mortality rate, survival and colonisation. Larvae are sensitive to changes in temperature and will be less active at 15°C than at room temperature or 37°C. They can be stored at 15°C before use, but it is important to allow larvae to warm to room temperature before infection, because lower temperatures can impair the effectiveness of the immune response, giving results that are not truly representative.

Larvae will continue to develop through their life cycle, so order the larvae only when you are ready to use them so that they do not begin to

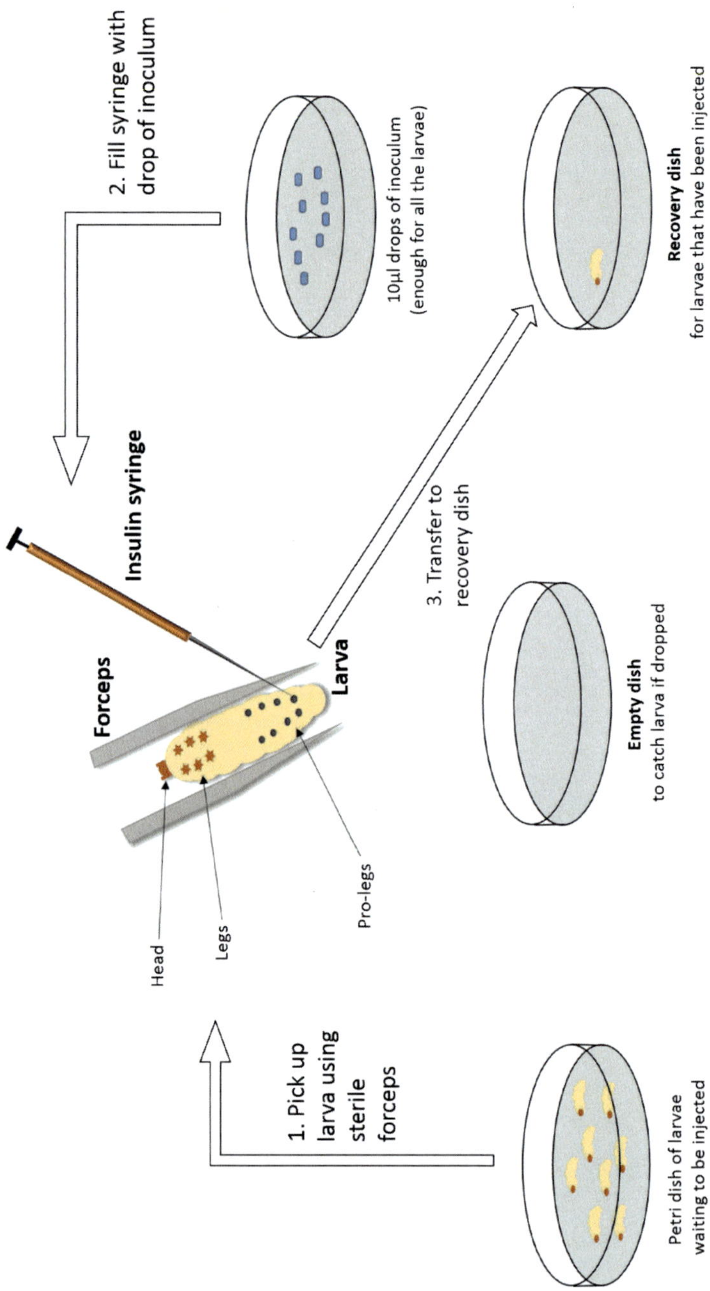

Figure 10.2 Schematic representation of how to set up a *Galleria mellonella* infection experiment.

pupate. *G. mellonella* usually arrive in plastic containers filled with sawdust and that have holes in the lids; they do not need to be fed. The containers tend to be small, so it is advisable to transfer them (and the sawdust) to a larger container with a lid that also has holes in it or is covered with foil that has been punctured to allow for air holes and secure the lid to the container with an elastic band or similar.

10.1.2.1 Experimental planning:

Typically, groups of 10 larvae are used in infection studies, which must include a control group for trauma and a control group for the vehicle (usually sterile phosphate-buffered saline [PBS]). Larvae weighing 0.2–0.3 g are chosen to allow for standardisation.

In addition to the two control groups, you will need to consider how many other test groups are required. If you are investigating the dose of bacteria it takes to kill all of the larvae over a given period, you will need to test a variety of different inoculum concentrations. To begin with, prepare a serial dilution of the bacteria of interest and enumerate the colony forming units (CFU) for each dilution. Then, choose a range of concentrations (which will then contain a known number of bacteria) to use to inoculate the larvae. The larvae will be infected with a volume of 10 μL, so adjust the CFU calculation (which will most likely be CFU mL^{-1}) so that you know how many bacteria are in the 10-μL dose.

When considering the rate of killing over time, a good time frame to begin with is 72 h, checking on the larvae and scoring the number of death every 24 h. This time frame is in keeping with most of those suggested for other experiments in this book, and so will provide consistent data.

Experimental setup (Fig. 10.2):

1. Prepare overnight cultures of the bacteria to be tested. The following day, prepare a serial dilution (to the known CFUs you determined earlier) in sterile PBS to encompass the range of doses to be tested. Keep the dilutions on ice.
2. Remove larvae from the 15°C incubator. Using one sterile Petri dish per test group, allocate 10 larvae (0.2–0.3 g per larva) to each test group. Label the plates with each test condition (including the trauma and vehicle controls) and leave them on the bench with the lids on.
3. Prepare a second set of sterile Petri dishes with the same group labels. These will be the recovery dishes for larvae after inoculation.
4. Prepare a third set of Petri dishes labelled with the same groups. These will be used to aliquot the inocula.

5. Select an appropriate number of disposable insulin syringes (one per larva) and keep them nearby, ensuring that you also have a sharps bin nearby for immediate disposal of used syringes. Alternatively, a fine Hamilton syringe can be used, but it must be cleaned out using ethanol between each dose.
6. Ensure you have a pair of forceps nearby with which to hold each larva, and place them in a sterile Petri dish with the lid off. This Petri dish will be useful if you accidently drop one of the larva when handling it, because it will contain it.

10.1.2.2 Experimental protocol:

(Figs 10.2 and 10.3)

1. Begin with the trauma group. Have the Petri dish containing the larvae on your left and the recovery dishes and inoculum dishes on your right. Make sure the forceps and spare dish are directly in front of you.
2. Pick up the first larva and hold it in the forceps. The larva must be on its back with its head pointing toward the closed end of the forceps. Hold it firmly but do not squeeze too hard; the larva should not be able to move.

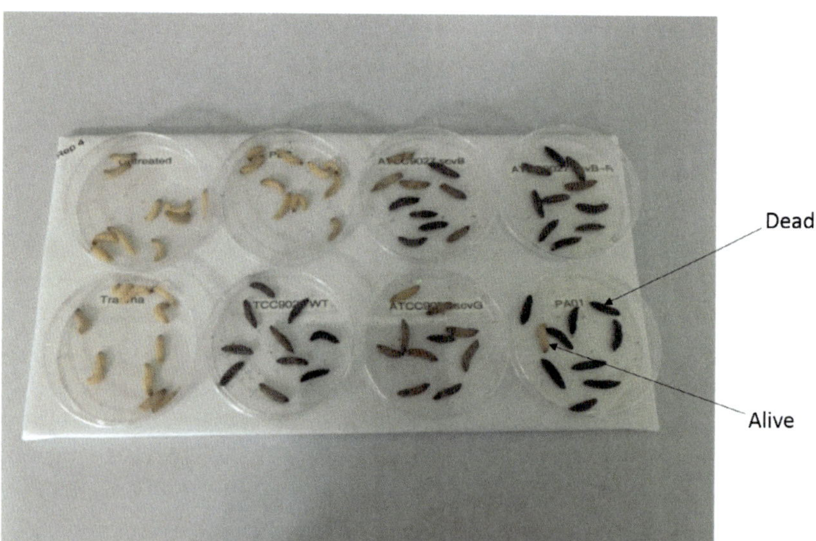

Figure 10.3 Completed *G. mellonella* infection experiment. Groups of 10 larvae are seen in each Petri dish. Cream-coloured larvae are alive and black larvae are scored as dead if unresponsive to touch with a pipette tip.

3. In your other hand, pick up the insulin syringe and insert the tip of the needle into the third left proleg of the larva. This does not need to be inserted deeply.
4. Dispose of the insulin syringe into the sharps bin and transfer the larva into the appropriate recovery dish. Do this for each larva.
5. For the PBS control and subsequent infection groups, the protocol is the same.
6. Next, do the same for the PBS control group.
7. Aliquot 10 × 10 μL drops of sterile PBS into a Petri dish. There is one drop per larva.
8. Pick up the first larva of the PBS control group.
9. With your other hand, draw up one of the 10-μL PBS drops into the insulin syringe (with no air).
10. Insert the tip of the needle into the first left proleg of the larva and inject the PBS.
11. Dispose of the insulin syringe into the sharps bin and transfer the larva into the appropriate recovery dish. Do this for each larva.
12. Repeat this procedure for each of the different inoculates.
13. Transfer the Petri dishes of larvae to a 37°C incubator and check them every 24 h for 72 h to see how many are dead and how many are alive.
14. Larvae that are infected might begin to turn black owing to the production of melanin; dead larvae will be completely black and will not respond when touched with the tip of a pipette.
15. At the end of the experiment, dispose of all larvae. If any larvae are still alive (the trauma and PBS controls should still be alive), transfer them to a freezer overnight and then dispose of all dead larvae (infected or not) by autoclave.

10.1.2.3 Testing an antimicrobial treatment:

Before testing antimicrobial treatments in this model, you must first determine the minimum bactericidal concentration (MIC) and minimum bactericidal concentration (MBC) (see Chapter 4). You can then test the antimicrobial at these doses compared with controls that do not receive the antimicrobial treatment. For these experiments, you will use just one dose of bacteria, preferably one that will kill half the population within 72h. That way, an increase or decrease in the numbers of larvae that survive when treated will be apparent.

The experiment is set up broadly as described previously, including a trauma and PBS control. Instead of having groups to test different inoculum doses, the test groups will use the MIC and the MBC (and possibly one dilution on either side of those values). The larvae will be infected as described; the ones receiving the treatment will be injected a second time, into the right proleg. The volume of the dose injected will be 10 μL. Petri dishes containing 10 μL aliquots of the treatment should be prepared as previously described. Survival of the larva should be monitored as per the killing assay. Alternatively, antimicrobials can be administered before infection to determine whether they offer prophylactic protection. Repeated dosing is possible, but if doing so, do not repeatedly inject into the same proleg, to minimise trauma to the larva.

It is possible to ascertain the reduction in bacterial load for infected larvae that have also received an antimicrobial treatment. The bacterial load for infected larvae that did not receive an antimicrobial should also be determined as a comparison. Bacteria can be recovered from larva that die as a result of the infection and those that survived. Bacteria are recovered from the haemolymph, and so all larvae should be killed by transferring them to a freezer overnight before collecting the haemolymph.

A protocol for determining the bacterial load in the haemolymph of *G. mellonella* is:

1. Remove larva from the freezer and allow to warm to room temperature.
2. Place the larva onto a Petri dish. Using a scalpel, make an incision between two segments near the tail.
3. Gently squeeze the larva to transfer the haemolymph into a 1.5-mL microcentrifuge tube and discard the larva by autoclaving.
4. Weigh an empty microcentrifuge tube and use this to determine the weight of the collected haemolymph in each tube.
5. Add an equal volume to weight of sterile PBS and mix
6. Prepare a serial dilution; enumerate as previously described.
7. It is advisable to use an agar that is selective for the bacteria you are recovering to eliminate any gut contaminants.

10.1.3 Kaplan–Meier analysis of data

Data generated from both the *C. elegans* and *G. mellonella* models can be assessed using Kaplan–Meier survival analysis. This uses the data collected to provide an indicator of survival in response to a given treatment; it is presented compared with a control. It is a commonly used method to present data obtained from *C. elegans* and *G. mellonella* infection models. This section describes how to process the data to produce a Kaplan–Meier plot.

First, you need to calculate the survival probability. This is done as shown in Fig. 10.4.

Time (h)	Total population	Total number that survived	KM survival probability
0	10	5	5/10=0.5
24	5	3	0.5x(3/5)=0.3
48	3	1	0.3x(1/3)=0.1
72	1	0	0.1x(0/1)=0

Figure 10.4 An example of the information that needs to be entered into a spreadsheet to calculate the Kaplan-Meier survival probability based on a *G. mellonella* infection model, with 10 larvae per group.

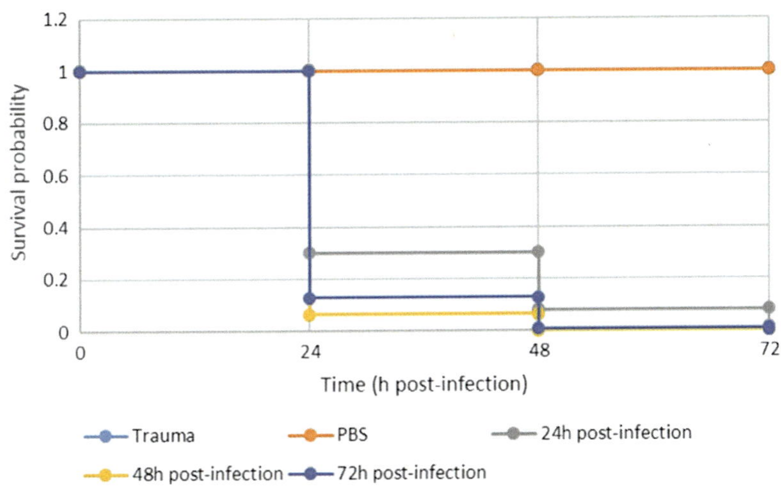

Figure 10.5 An example of a Kaplan Meier curve showing the survival of infected larvae, and including all appropriate controls.

10.1.4 Example of information to be entered into spreadsheet to calculate Kaplan–Meier survival probability based on *G. mellonella* infection model, with 10 larvae per group

This will give a survival probability score. These values are plotted on the y-axis. Time (in days or hours) is plotted along the x-axis. The type of graph that is plotted is a step-chart, and to plot this type of graph you will need to input two data points per time point, which will produce a graph like the one shown in Fig. 10.5. This type of graph does not require specialist statistical software and can be done using packages such Excel or Numbers.

10.2 Summary

Having gathered a robust set of in vitro data, these two infection models will enable you to take the next step toward understanding pathogenicity, virulence and the effectiveness of potential antimicrobial treatments. Although there are limitations to these models, they are an excellent starting point for in vivo studies and an excellent way to conclude the comprehensive in vitro analyses described in this book.

10.3 Notes page

Record observations and notes here.

References

Cisar, J.O., Kolenbrander, P.E., McIntyre, F.C., 1979. Specificity of co-aggregation reactions between human oral *Streptococci* and strains of *Actinomyces viscosus* and *Actinomyces naeslundii*. Infection and Immunity 24, 742–752.

Duckworth, P.F., Rowlands, R.S., Barbour, M.E., Maddocks, S.E., 2018. A novel flow system to establish experimental biofilms for modelling chronic wound infection and testing the efficacy of wound dressings. Microbiological Research 215, 141–147.

European Committee for Antimicrobial Susceptibility Testing (EUCAST) of the European Society of Clinical Microbiology and Infectious Diseases (ESCMID), September 2000. Terminology relating to methods for the determination of susceptibility of bacteria to antimicrobial agents. Clinical Microbiology and Infections 6 (9), 503–508.

Garsin, D.A., Sifri, C.D., Mylonakis, E., Qin, X., Singh, K.V., Murray, B.E., Calderwood, S.B., Ausubel, F.M., 2001. A simple model host for identifying Gram-positive virulence factors. Proceedings of the National Academy of Sciences 98, 10892–10897.

Goeres, D.M., Hamilton, M.A., Beck, N.A., Buckingham-Meyer, K., Hilyard, J.D., Loetterle, L.R., Lorenz, L.A., Walker, D.K., Stewart, P.S., 2009. A method for growing biofilm under low shear at the air-liquid interface using the drip flow biofilm reactor. Nature Protocols 4, 783–788.

Hall, B.M., Ma, C., Liang, P., Singh, K.K., 2009. Fluctuation analysis calculator: a web tool for the determination of mutation rate using Luria-Delbruck fluctuation analysis. Bioinformatics 25, 1564–1565.

Konrat, K., Schwebke, I., Laue, M., Dittmann, C., Levin, K., Andrich, R., et al., 2016. The bead assay for biofilms: a quick, easy and robust method for testing disinfectants. PLoS One 11, e0157663.

Laws, T.R., Harding, S.V., Smith, M.P., Atkins, T.P., Titball, R.W., 2004. Age influences resistance of *Caenorhabditis elegans* to killing by pathogenic bacteria. FEMS Microbiology Letters 234, 281–287.

Lomolino, A., Di Perro, G., Lante, A., 2012. A quantitative fluorescence-based lipase assay. Food Technology and Biotechnology 50, 479–482.

Maddocks, S.E., Jenkins, R.E., 2017. Understanding PCR. Elsevier, London, pp. 1–87.

Marques, M.R.C., Loebenberg, R., Almukainzi, M., 2011. Simulated biological fluids with possible application to dissolution testing. Dissolution Technologies 18, 15–28.

Odds, F.C., July 1, 2003. Synergy, antagonism, and what the chequerboard puts between them. Journal of Antimicrobial Chemotherapy 52 (1), 1.

Stiernagle, T., 1999. WormBook: The Online Review of *C. elegans* Biology. http://www.wormbook.org/chapters/www_strainmaintain/strainmaintain.html.

Tolker-Nielsen, T., Sternberg, C., 2011. Growing and analysing biofilms in flow chambers. Current Protocols in Microbiology (Chapter 1):Unit 1B.2.

Unno, M., Cho, O., Sugita, T., 2017. Inhibition of *Propionibacterium acne* lipase activity by the antifungal agent ketoconazole. Microbiology and Immunology 61, 42–44.

Webpages

American Type Culture Collection. https://www.lgcstandardsatcc.org/Products/Cells_and_Microorganisms/Bacteria.aspx?geo_country=gb.

BD Diagnostics. http://www.bd.com/europe/.

BioMerieux. https://www.biomerieux.co.uk.

Biorad DGGE System. http://www.bio-rad.com/webroot/web/pdf/lsr/literature/M1709080C.pdf.

Chromagar. http://www.chromagar.com/.
Clinical and Laboratory Standards Institute. https://clsi.org/.
Eurofins. www.eurofins.co.uk.
Fisher Scientific. https://www.fishersci.co.uk.
Hardy Diagnostics. http://hardydiagnostics.com/.
Health and Safety Executive. http://www.hse.gov.uk/index.htm.
Liofilchem. https://www.liofilchem.com/en/.
Merck. https://www.sigmaaldrich.com/united-kingdom.html.
National Centre for Biotechnology Information. https://www.ncbi.nlm.nih.gov.
Oxoid. http://www.oxoid.com/UK/blue/search/results.asp.
Prosource Scientific. http://www.psscientific.com/default.aspx.
Public Heath England National Collection of Type Culture. (https://www.phe-culturecollections.org.uk/collections/nctc.aspx).
SigmaAldritch. https://www.sigmaaldrich.com/united-kingdom.html.
The *Caenorhabditis* Genetic Centre. https://cbs.umn.edu/cgc/home.
The European Committee on Antimicrobial Susceptibility Testing. http://www.eucast.org/.
The Pseudomonas Genome Database. www.pseudomonas.com.
ThermoFisher Scientific. www.thermofisher.com.
Tools for Handling *C. elegans*. https://cbs.umn.edu/cgc/home.
VWR. https://uk.vwr.com/store/;jsessionid=VRhV16QI9jHOBACYAOX96VVH.estore5b.

Index

Note: 'Page numbers followed by "f" indicate figures, "t" indicates tables.'

Made in the USA
Las Vegas, NV
17 April 2025

21015321R10138